Chemical and Biochemical Catalysis for
Next Generation Biofuels

Chemical and Biochemical Catalysis for Next Generation Biofuels

Edited by

Blake Simmons
Joint BioEnergy Institute, Sandia National Laboratories, Emeryville, CA, USA

RSCPublishing

RSC Energy and Environment Series No. 4

ISBN: 978-1-84973-030-3
ISSN: 2044-0774

A catalogue record for this book is available from the British Library

Published by The Royal Society of Chemistry,
Thomas Graham House, Science Park, Milton Road,
Cambridge CB4 0WF, UK

Registered Charity Number 207890

For further information see our web site at www.rsc.org

Printed in Great Britain by CPI Group (UK) Ltd, Croydon, CR0 4YY

Contents

RSC Energy and Environment Series No. 4
Chemical and Biochemical Catalysis for Next Generation Biofuels
Edited by Blake Simmons
© Royal Society of Chemistry 2011
Published by the Royal Society of Chemistry, www.rsc.org

Preface

The development and realization of renewable and sustainable lignocelluosic biofuels is an area that is receiving worldwide attention and investment. Despite several decades of research, there remain significant challenges that need to be overcome before these biofuels can be produced in large volumes at cost-competitive prices. Some of the most significant obstacles are found in the development of efficient catalytic systems that can convert the biomass into intermediates (e.g. sugars) and/or biofuels. The price of these catalysts, be they biological, thermochemical, or chemical in nature, represents one of the largest costs in the conversion process. There exist a significant number of potential catalytic schemes available, and each has certain disadvantages and advantages that must be considered when selecting the appropriate system.

This book will present a general but substantial review of the most promising of these catalytic processes, and will cover the spectrum of biomass pretreatment, enzymes, chemical catalysts, and hybrid approaches of hydrolyzing biomass into fermentable sugars and/or directly into biofuels. The book starts with a comprehensive evaluation of feedstocks in terms of availability, sustainability, and land cover type. The subsequent chapters focus on detailed assessment around biomass conversion technologies, including biomass pretreatment, chemical catalysis, thermochemical conversion, and bioinspired catalysts.

RSC Energy and Environment Series No. 4
Chemical and Biochemical Catalysis for Next Generation Biofuels
Edited by Blake Simmons
© Royal Society of Chemistry 2011
Published by the Royal Society of Chemistry, www.rsc.org

Introduction

BLAKE A. SIMMONS

Joint BioEnergy Institute, Physical Biosciences Division, Lawrence Berkeley National Laboratory, Emeryville, CA; Biofuels and Biomaterials Science and Technology, Sandia National Laboratories, Livermore, CA

The development of advanced biofuels, defined here as those that are derived from non-food sources, capable of displacing a significant amount of petroleum within the global transportation sector has quickly become a topic of significant interest. The primary drivers for this effort are found in two areas: (1) concerns over energy security related to finite sources of fossil fuels, and (2) the environmental risks associated with unabated carbon emissions that are linked to global warming. While there are significant efforts underway in the fields of renewable energy for electricity production (e.g. wind, solar and geothermal), more than half of the current energy consumption of the planet is currently met with the consumption of liquid fuels, with over 20% of current global carbon emissions in 2008 generated by the transportation sector alone (Figure 1.1). The significant price fluctuations observed in the petroleum ($32-147/barrel) and natural gas ($4-13/1000ft^3) markets over the past few years have had dramatic effects in all aspects of commerce, and increase concerns over resource availability and supply stability.

Recent estimates by the International Energy Agency (IEA) calculate that global demand for transport will increase by 45% by 2030, placing even further strains on a system that has reached the upper limit in terms of production and further increase carbon emissions.[1] These results underscore the need for the realization and rapid commercialization of scalable and cost-effective means of

RSC Energy and Environment Series No. 4
Chemical and Biochemical Catalysis for Next Generation Biofuels
Edited by Blake Simmons
© Royal Society of Chemistry 2011
Published by the Royal Society of Chemistry, www.rsc.org

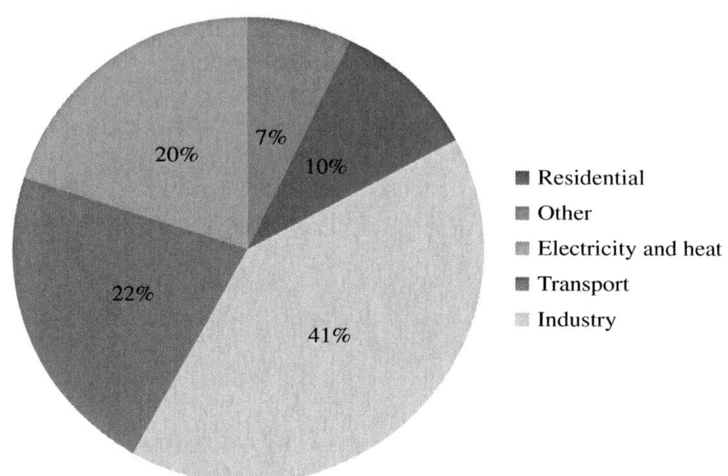

Figure 1.1 Distribution of global carbon emissions as a function of energy sector. (Adapted from IEA Report: CO_2 Emissions from Fuel Combustion-Highlights, 2010, pg. 9.)

generating low-carbon fuels to meet the growing global energy demand in a sustainable fashion.

To meet this challenge, the US federal government, several state governments, and numerous governments worldwide are strongly committed to displacing fossil fuels with renewable, low carbon fuels produced from biomass. For instance, the US federal government has set a target of displacing 36 billion gallons of current US petroleum consumption within the transportation sector by 2022 under the Renewable Fuel Standard (RFS2) legislation. With a production cap of 15 billion gallons per year placed on corn ethanol, this leaves a gap of 21 billion gallons per year that must be met by other sources and conversion technologies. With total fossil fuel consumption within this sector currently running at levels of ~ 200 billion gallons per year in the United States, this requires the development of a significant commercial infrastructure capable of producing approximately this level of non-starch biofuels per year over a very short timeframe. The European Union, China, Australia and New Zealand have also established targets for biofuel production.

The development of these advanced biofuels is not without controversy, and the recent furor over the "food vs. fuel" debate[2] has highlighted the need for the development of sustainability metrics within the context of global food and energy supplies. The annual global primary production of biomass, or lignocellulose, is equivalent to the 4500 EJ of solar energy captured each year.[3] It is estimated that a sustainable bioenergy supply of 270 EJ can meet $\sim 50\%$ of the world's total primary energy demand. In terms of total biomass availability, this amount of bioenergy can be achieved by only using $\sim 6\%$ of the annual global primary production of biomass. As with all terrestrial systems of production, the ultimate potential for bioenergy depends to great extent on the land available for production. Currently, the amount of land devoted to growing energy crops for biomass fuels is calculated to represent only 0.19% of the world's total land area and only 0.5–1.7% of global agricultural land.[4] If we establish an upper limit of the total global bioenergy production potential in 2050 of 1135 EJ, out of a total global energy demand of 1041 EJ (source: EIA, 2011), it is theoretically possible that the total global energy demand can be met on a renewable basis without affecting the global production of food crops.[5]

There exist multiple pathways that have been hypothesized as effective means of converting biomass into biofuels, biopower, and co-products. These include biochemical, chemical, thermochemical, and hybrid conversion routes that follow multiple pathways for the production of fuels and chemicals (Figure 1.2). All of the technologies to date have advantages and disadvantages that must be taken into consideration, and all of them are currently deployed at some scale throughout the world. More importantly, the logistics of the local environment in which a proposed biorefinery/biopower conversion unit is built must be considered when identifying the most appropriate conversion technology, or combination of routes, for that region. Water availability, biomass availability, advanced land management practices, fertilizer demand, harvesting, storage, and distribution systems are all critical aspects in the successful operation of any sustainable biorefinery lifecycle that must be considered at a systems level.

In addition to the sustainability criterion, cost is another major factor that must be addressed before advanced biofuels can reach significant levels of

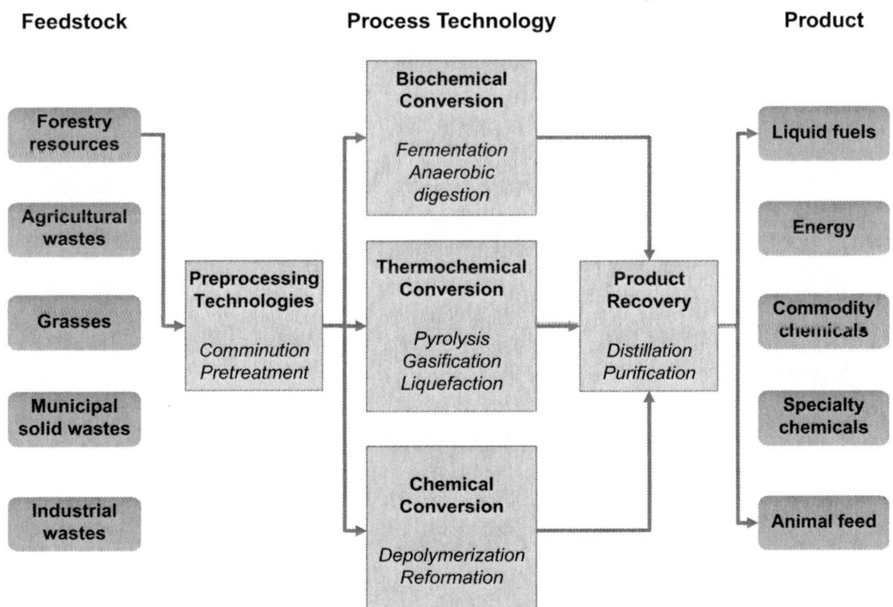

Figure 1.2 Schematic depiction of potential biomass conversion pathways.

production. The cost estimates for these advanced biofuels varies significantly within the scientific and commercial literature. A recent report by the National Academy of Sciences estimated that cellulosic ethanol produced by biochemical conversion is equivalent to \$115/bbl of gasoline and that biomass-to-liquid biofuels produced by thermochemical conversion are equivalent to \$140/bbl.[6] These estimates are predicted future prices and are highly dependent on assumptions of feedstock cost, conversion efficiency, and type of fuel produced, and as such there remain significant opportunities around biomass conversion technologies that could significantly reduce the cost of production.

One of the most significant obstacles today is the lack of affordable catalytic systems that can efficiently convert the biomass into desired intermediates (e.g. sugar) and/or products (e.g. fuels). The price of these catalysts, be they biological, thermochemical, or chemical in nature, represent one of the largest costs in the conversion process. For example, in the biochemical conversion route, it has been calculated that the cost of deconstructing the biomass (biomass→sugars) is second only to the cost of the feedstock, and represents the single biggest opportunity for cost savings within the biorefinery context. Similar cost pressures exist for the thermochemical conversion route, where catalysts must be able to tolerate complex heterogeneous feeds while maintaining high selectivity and high yields at low cost. There are a number of catalytic schemes, each with their own advantages and disadvantages, currently under development. This book presents a general yet substantial review from subject matter experts of the most promising processes within the spectrum of biomass pretreatment, enzymes, chemical catalysts, thermochemistry and hybrid approaches of converting biomass into intermediates and fuels, and is aimed to inform the reader on a wide range of topics.

References

1. IEA, *CO$_2$ Emissions from Fuel Combustion - Highlights*, 2010, 130 pp.
2. A. O. Converse, Renewable energy in the United States: is there enough land? Applied Biochemistry and Biotechnology, 2007, **137–140**, 11–624.
3. R. H. Sims, *Bioenergy Options for a Cleaner Environment: In Developed and Developing Countries*, 2004, Elsevier Ltd., Oxford, UK, 184 pp.
4. M. Hoogwijk, A. Faaij, R. Van den Broek, G. Berndes, D. Gielen and W. Turkenburg, Exploration of the ranges of the global potential of biomass for energy, *Biomass and Bioenergy*, 2003, **25**(2), 119–133.
5. E. M. W. Smeets, A. P. C. Faaij, I. M. Lewandowski and W. C. Turkenburg, A bottom-up assessment and review of global bio-energy potentials to 2050, *Progress in Energy and Combustion Science*, 2007, **33**(1), 56–106.
6. National Academy of Sciences, *Overview and Summary of America's Energy Future: Technology and Transformation*, 2010, 58 pp.

Biomass Availability and Sustainability for Biofuels

DOMINIQUE LOQUÉ, AYMERICK EUDES
AND FAN YANG

Feedstocks Division, Joint BioEnergy Institute, Physical Biosciences
Division, Lawrence Berkeley National Laboratory, Berkeley, CA

2.1 Introduction

On earth, only 29.2% of the surface is land (149 million km^2) and the rest is covered ocean. From this 29.2%, only 59.5% is considered as biologically productive land (88.6 million km^2) which corresponds to forests (39.3 million km^2) or agricultural areas (49.3 million km^2). Biologically productive land corresponds to land that supports human demands for food, fiber, and timber for infrastructure and energy (FAO definition). The other 40.5% of lands, considered as non-productive lands, have a very low or no primary productivity since they are covered by ice, water, or constructions, or they are located under extreme climate conditions (cold, dry, or arid). The productive lands are divided into several biomes, primarily classified according to the vegetation types and productivity,[1] which are dictated by the climate and human accessibility. In order to define which lands can be transformed as bioenergy lands (biofuel lands), an evaluation of most of the primary lands has to be conducted and will be presented in this section.

RSC Energy and Environment Series No. 4
Chemical and Biochemical Catalysis for Next Generation Biofuels
Edited by Blake Simmons
© Royal Society of Chemistry 2011
Published by the Royal Society of Chemistry, www.rsc.org

2.2   General Land Types

2.2.1   Forest Lands

Forests cover approximately 39.3 million km^2 and are divided into three main types: boreal, tropical, and temperate (Table 2.1). Boreal forests represent 13 million km^2, 33% of the total forest, and correspond to forest growing in cold areas (yearly average temperature $+5$ to $-5\,^{\circ}$C) and a short growing season with an aboveground biomass accumulation of 2.3 t/ha/year.[2] They are mainly found in the northern part of the northern hemisphere and in some mountains at high altitudes. Coniferous trees are the dominant species, also called 'evergreen'.

The temperate forests cover 9.8 million km^2, 25% of total forest,[2] and are found in a more moderate climate and in both hemispheres. The diversity of tree species is much larger than the boreal forest and varies significantly between both hemispheres. The dominant species eucalyptus, Nothofagus, Araucaria, and Podocarpus are predominant in the southern hemisphere and pine, sequoia, oak, maple, and birch are preferentially in the northern hemisphere. The temperate forests cover a smaller surface than the boreal or tropical forests. They are mainly found on low quality soils (sandy, rocky, etc.), on poorly accessible areas or in isolated areas, which correspond to lands that are usually classified as non-suitable for farming. This is explained by a large deforestation during the past centuries that were made to increase agricultural area and was responsible of the conversion of the best forestlands into croplands. The approximate aboveground biomass accumulation of temperate forests is estimated at 9.5 t/ha/year.[2]

Table 2.1   Summary of submerged-lands.

		Area			Productivity
		(M km^2)	*%*	*%*	*(t DM/ha/y)*
Forest lands	boreal	13	33		2.3
	tropical	9.8	25		9.5
	temperate	16.5	42		14
		39.3		**26.4**	
Agricultural lands	very suitable	13.5	27		18–22
	suitable	15.1	31		13–17
	moderately suitable	7.9	16		10–12
	marginally suitable	5.4	11		4–7
	non-suitable	7.4	15		<2
		49.3		**33.1**	
"Non-biologically" productive lands	Desert	31.5	52		1.5
	Tundra	5.6	9		0.8
	Rest (urban area, rivers, glacier…)	23.3	39		
		60.4		**40.5**	
Submerged lands		**149**			

(Data extracted from: http://faostat.fao.org with 2008 as reference year, Terrestrial Global Productivity[2], and several other resources.[5,6,10])

Tropical forests represent almost 50% of the world's forest and cover 16.5 million km^2, 11% of the available land.[2] They are mainly found under the wet tropical climate located around the equatorial regions with a monthly temperature average above 18 °C and a monthly rain above 100 mm monthly.[3] These giant forests contain the largest species diversity that has been estimated at over >65 000.[4] Tropical forest is the biome that has the highest biomass productivity, with a total aboveground biomass accumulation of 14 t/ha/year approximately.[2]

2.2.2 Agricultural Lands

Forests cover 45.7% of biologically active lands at 49.3 million km^2, and correspond to lands that have an agricultural potential and are mainly represented by croplands and pastures (Table 2.1). It is estimated that 31% of these lands (15.5 million km^2) are used for food and feed production. From the 38.8 million km^2 designated as pastures, it is estimated that 70% (26.4 million km^2) has the potential to be converted to croplands since they would be suitable for rain-fed agriculture.[5]

From the 49.3 million km^2 of agricultural lands, 41.9 million km^2 of lands are considered to be suitable lands for crop production and are divided as 13.5 million km^2 as very suitable crop lands with a productivity of 18–22 t DM/ha/year, 15.1 million km^2 as suitable crop lands with a productivity of 13–17 t DM/ha/year, 7.9 million km^2 as moderately suitable crop lands with a productivity of 10–12 t DM/ha/year, and 5.4 million km^2 as marginally suitable crop lands with a productivity of 4–7 t DM/ha/year.[6] These lands were classified according to their potential food-crop productivity for rice, wheat, corn, or soybean. Interestingly, most of these potentially available lands would come from developing countries and the very suitable, suitable, moderately suitable, and marginally suitable land correspond to 80–100%, 60–80%, 40–60% and 20–40% respectively of the maximal attainable annual biomass.[5] The estimation of productivity is based on *Miscanthus* yields produced in Eastern Europe.[6] The conversion of pastures into croplands needs to be conducted with caution to avoid irreversible losses since it is already estimated to 3.1 million km^2 of lands have been severely or irreversibly degraded since 1945 due to severe chemical degradation and/or erosion.[7,8] It is also estimated that 38% (5.9 million km^2) of actual croplands are suffering from degradation and consequently have lost part of their respective productivity potential.[8] It is also important to note that from the 15.5 million km^2 of rain-fed croplands, 2.9 million km^2 are assisted by irrigation, which, over the long term, is partially responsible for land degradation.[9]

2.2.3 Desert Lands

A third large area, which represents 21.1% (31.5 million km^2)[10] of available lands, corresponds to the desert ecosystems that are not part of the biologically productive lands so far (Table 2.1). Deserts are usually found in areas that have the lowest and most irregular yearly rainfalls (below 250mm/y), and, consequently, have very low biomass productivity, typically below 1.5 t/ha/year.[2] Arid, semi-arid, and coastal deserts are constantly expanding due to desertification, a physical process which is very frequently caused by human activity. Desertification in arid areas is not a phenomena that is only associated to a lack of precipitation, it can be indirectly caused by reduction of the vegetative cover due to nutrient deficiency, mainly nitrogen and phosphorus deficiencies, overgrazing, erosion, and salt stresses.[11–13] In fact, the decline of the vegetative cover reduces the organic matter that can retain moisture in the soil and nutrients, but also increases soil evaporation and soil erosion. The main plant species found in the desert are short life cycle annual plants and drought stress-tolerant plants such as shrubs, succulents, and few trees.

2.2.4 Tundra Lands

The last major biome on earth, tundra, is the coldest ecosystem, and together with the desert biomes have the lowest biomass productivity on earth. The small amount of rainfall and limited number of days with a temperature that goes above 0 °C renders the area not suited for trees. The plant species able to grow under these extreme conditions are mainly small numbers of specialized mosses, grasses, lichens, and shrubs.[14] Growth primarily occurs during the continuous daylight that occurs in summer after the snow is melted and the ground defrosted. This biome is mainly found near the polar circle on the border of the boreal forests. This biome covers about 5.6 million km^2.[2] Even if the biomass productivity $(0.8 \text{ t/ha/y})^2$ is as low as or even lower than the desert ecosystems, the carbon stored in the underground is very important due to the low biomass degradation rate.[15,16]

2.3 Potential Bioenergy Feedstock Lands

Lands that could be potentially suitable for biomass production for biofuels need to be lands that have enough productivity to render this approach financially sustainable. The higher the land productivity will be, the lower the quantity of land surface required to meet the needs and lower the ecologically impact will be. However, the most productive lands are either the tropical forests that should be protected or the best croplands already used for human agriculture. Lands that are still productive but non-sustainable for food production, are called marginal lands. They would potentially represent 28 million km^2. Switchgrass, one of the potential bioenergy crops, has a yield varying between 5.6 and 18 t DM/ha/y, depending on the species and geography, but a majority of them yield above 10 t DM/ha/y.[17] The growth of bioenergy crops, such as switchgrass, on 28 million km^2 yielding at 10 t DM/ha/y (low-end yield), would generate 28 billion t/y of dry biomass, which represents approximately 518×10^{18} J/y (2.8 billion ha $\times$ 10 t/ha $\times$ 18.5 billion J/t DM).

The amount of energy that could be potentially produced on the lower quality of biologically active lands is just above the actual worldwide energy consumption estimated at 498×10^{18} J in 2006 (http://www.eia.doe.gov/oiaf/ieo/world.html) or 393×10^{18} J (80.1% of total energy: gas, oil, and coal) from fossil fuels.[18] The amount of arable lands left over corresponds to the topsoil lands and represents double of what is currently used for food production. The production of energy from biomass could be further extended since the proposed land does not include any forest lands, in fact doubling the surface of the currently managed forest lands, estimated at 3% (1.18 million km^2) and corresponding of boreal and temperate would have the potential to generate more biomass without competing with food crop lands.

The proposed bioenergy croplands could be reduced if the selection of the bioenergy crops is done according to the available land quality and if breeding programs are developed on dedicated bioenergy crop species. For example, in the paper industry significant improvements have been achieved to improve pulping efficiency and yield, and are mainly attributed to an increase of timber yield and quality which have been addressed by genetic improvement and better forest management, and finally by better conversion processes.[19–23] This suggests that similar improvements of energy crops and conversion techniques have the potential to reduce the land area required to replace the 393×10^{18} J produced from fossil fuels[18] and/or to compensate for the constant augmentation of worldwide energy consumption. In addition, the lands proposed as bioenergy crop lands could be reduced if abandoned lands and non-biologically-active lands (FAO definition) are targeted for bioenergy production. Additionally, there is a huge amount of unused biomass generated on various lands; forest lands, agricultural lands, and city lands, available for conversion.

2.4 Bioenergy Feedstocks

Plants are divided into three main groups according to their longevity; annual, biannual, and perennial. These groups have common and complementary agronomical properties that are fully used by farmers and landscape managers. With the vision of land sustainability and low input, (energy, time, and fertilizer), in general, perennial plants have a serious advantage. They have the ability to grow on marginal lands and to avoid land degradation especially caused by erosion (water, wind, and water) since they can fix the soil and bring a yearly land coverage in contrast to annual plants that have a dead time of land coverage between the end of the fruiting time and the newly developed plant. The planting of perennial species on degraded lands or abandoned farmlands can be converted into biofuels sustainably as they represent feedstocks produced with little or no competition with food production since these lands are usually improper or unsustainable for food crop production. The use of perennial plants that require little input in terms of fertilizer and can be grown on rain fed water, have the benefits of generating a lower greenhouse gas emissions than traditional fossil fuels. However, double crops and mixed cropping systems are also considered for biofuel production since some of them have the potential to keep a permanent vegetative cover to mitigate the potential risk of land degradation due to erosion.

Similarly to any other crop, high biomass yields obtained from dedicated bioenergy crops is desirable. This can be achieved through the improvement of several traits to maximize photosynthesis. These include improving light radiation interception, use efficiency, and CO_2 fixation efficiency.[24,25] In particular, leaves could be more efficient at capturing light and CO_2 to show higher photosynthesis rates. Plants able to grow under high density and developing canopies with low extinction coefficients are considered to intercept more light radiation. Additionally, the prolongation of the light capture time and CO_2 fixation could be achieved by the extension of the vegetative growth using plants with a delay of senescence.[26–29]

Bioenergy crops should also meet the traits usually favored in traditional agriculture such as cold and drought tolerance, resistance to lodging, efficient and flexible in terms of water use, disease and pest resistance, and high nitrogen-use efficiency. Nitrogen is a very expensive nutrient when it needs to be manually supplied since nitrogen fertilizers require a lot of energy to be made and intensive supply create lot of environmental disasters. Also, plants providing insufficient biomass for harvest during their first year of culture should show winter standing capacity, as it would improve the light capture efficiency during the following years.

Furthermore, in order to ensure environmental and financial sustainability of dedicated bioenergy crops, several plant traits that would minimize fossil fuel inputs and nutrient depletion have to be considered. They include efficient nutrient recycling to the roots and remobilization, optimal root/shoot partitioning, low nitrogen requirements, and efficient use of water. Bioenergy-crops adapted to wastewater supply, polluted, and salty conditions represent attractive resources, as they would not compete with the water dedicated for food crops. Bioenergy crops able to grow on poor soils or marginal lands would not occupy lands required for food crops and offer the opportunity of restoring eroded lands by sequestering additional soil carbon. Bioenergy crops should be non-invasive or sterile in order to control their spatial distribution to avoid ecological catastrophes.

Alternatively, multiple-use scenarios for which crops that produce food and crop residues for biofuel can be considered. However, the amount of crop residue that can be removed from the soil has to be determined carefully in order to not impact soil and environmental quality as well as future crop yields grown on those lands.[30]

Sustainability of bioenergy crops will increase if they are adapted to farming systems (i.e. equipment, transport) and suitable crop rotation patterns, harvesting, and storage. Consequently, plants should be easily removed from the soil and grown from seeds, should generate straight with upright stems, have low moisture content and high sugar density, and finally should be resistant to microbial breakdown during post-harvest. Plants that provide enough biomass during the first year of planting would be desirable to optimize the harvest and productivity.

After harvesting, multiple conversion methods and flexible processing of the biomass should be employed to maximize the value of the feedstock. Plants such as sugarcane represent feedstocks requiring low or no pretreatment to harvest soluble sugars as opposed to lignocellulosic biomass, which requires pretreatment and hydrolysis processes prior to fermentation of the released sugars.

Several plant species are already considered as good candidates since they fulfill part of requirement described above to be a good bioenergy crop (Table 2.2). These crops can be categorized as 'energy cane' such as sugarcane, perennial warm-season grasses (WSGs) such as *Miscanthus*, switchgrass, and sorghum, and short-rotation woody crops (SRWCs) such as poplar, willows, and eucalyptus. WSGs have high biomass yield, deep-root system, rapid growth, low-maintenance, greater adaptability, and higher drought tolerance as compared with other common grass species.[31] Growing perennial WSGs does not require intensive cultivation or soil disturbance after establishment, which potentially offers soil and environmental improvements. Because WSGs enhance nutrient cycling and storage, and deep rooting system, they may require lesser amounts of fertilizers and may suffer less of water limitation than annual food crops. The magnitude of benefits of growing WSGs will depend,

Table 2.2   Properties of different plant species considered for bioenergy-crops.

	Sugarcane	Miscanthus	*Sorghum*	Switchgrass	*Short-rotation trees (Poplar, Eucalyptus)*
Efficient photosynthesis	■	■	■	■	
Long canopy duration	■	■	■	■	■
Nutrients recycled to roots		■		■	■
Low crop inputs		■	■	■	■
Low fossil fuel inputs		■		■	■
Adapted to marginal land		■	■	■	■
Minimal pests/plant diseases		■		■	■
Non-invasive or sterile	■	■	■	■	■
Easily removed	■	■	■	■	■
Winter standing	■	■		■	■
High water-use efficiency	■	■	■	■	■
Planted by seed			■	■	■
Harvest first year			■		
Adaptation to farming systems	■	■	■	■	
Genetic/biotechnology		■		■	■
Low pretreatment cost to obtain sugars	■				

however, on the culture management; harvest frequency, cutting height, and energy density.[31]

SRWC can be harvested faster after establishment compared to traditional forestry,[32] and economical losses and recovery time from wildfire will also be reduced. Plantations of SRWCs generate high input of leaf and root litter, coupled with reduced soil disturbance, which benefit soil properties by minimizing crusting. Growing SRWCs such as *Salix* and *Populus* spp. influences soil aggregate stability, aggregate strength, soil porosity, soil water retention, and aeration. The SRWCs can also improve soil water retention over cultivated soils due to their greater soil organic matter concentration.[33] Greater accumulation of soil organic carbon under SRWCs will occur when trees are grown in marginal and eroded lands rather than in croplands or natural forests.[34] SRWC produces more biomass over longer periods of time (12–15 years) and a higher biomass/sugar density compared to WSGs, thus representing more manageable reserves of feedstock supplies that do not require frequent harvest and storage processes.

Growing dedicated plants with the bioenergy crop properties (see above and Table 2.2) would minimize the use of the land dedicated and potentially available for food crops and would have a lower impact on the natural diversity found on marginal lands. Additionally, the use of degraded or polluted lands not suitable for food crops to produce biofuel crops could be restored as good land and potentially become suitable for food production. Perennial energy crops (trees and grasses) have the ability to remobilize nutrients in their root system during senescence, offering a source of minerals and organic matter in the soil after harvesting the aboveground biomass. Alternatively, depletion of toxic compounds from the soil could be achieved when growing plants capable of extracting and accumulating high levels of pollutants such as heavy metals in their aboveground tissues. The ability of WSGs and SRWCs to grow on lands with deteriorated soil conditions that are not appropriate for growing conventional crops suggests that some agricultural lands can be used to produce biofuel feedstocks and become profitable.

Feedstocks subtracted from fertile lands (i.e. lands already occupied with food crops) could be the biomass produced from double crops and mixed cropping systems. For example, bioenergy crops grown and harvested before the sowing and growing seasons of conventional food is an example of land-use options with potential to produce biofuel feedstocks under good management without decreasing food production and without clearing wild lands.[35] It could have the advantage of creating a plant cover during intercropping which would have the advantage of reducing soil erosion. Mixed cropping systems in which food and energy crops are grown simultaneously is another possibility.[36] Finally, plants such as alfalfa are considered as a feedstock in a 'multiple-use' scenario – the fractionation of alfalfa into a leaf fraction for high quality forage and a stem fraction for cellulosic biofuel.[37]

Biotechnology and the design of genetically improved plants would considerably upgrade bioenergy crops sustainability. Such approach is aiming at improving biomass properties in terms of growth efficiency, quantity, and quality (i.e. biomass degradability and added value). Example of poplar genetic transformations were first published 20 years ago and various poplar species have been studied for genetic engineering purposes, with for example the deregulation of specific genes involved in biomass quality.[38] Switchgrass transformation protocols are also now well established and open an avenue for genetic improvement.[39] The potential genetic transformation of willows and *Miscanthus* is less advanced, but research on embryogenic suspension culture and the transformation of callus tissue of *Miscanthus* via microprojectile bombardment, suggest that the generation of transgenic *Miscanthus* clones should be possible in the near future. In particular, plant height is one of the most important biomass yield components for which key controlling genes and pathways have recently been identified. Manipulating plant height in model species using genetic engineering was recently achieved, and several genomic regions that influence plant height in maize and sorghum have been identified.[40]

2.5 Degraded and Non-productive Lands

The international Soil Reference and Information Center (ISRIC) and the United Nations Environment Programme (UNEP) published a report on the degradation of agricultural lands due to erosion in a document entitled 'World Map of the Status of Human-Induced Soil Degradation'.[41] Land degradation is a natural process, but it is very often accelerated by human activities such as over-grazing, over-cultivation, over-irrigation, deforestation, and industrial pollution. This degradation can also be caused by the accumulation of organic pollutants, heavy metals, or salt, water, wind, or drought stresses, and will commonly be associated with erosion processes. In the GLASOD report[41] the authors claim that a total of 19.64 million km^2 were degraded worldwide in 1991. Water erosion apparently affects 10.94 million km^2 (56% of the total area suffering degradation). Wind erosion affects 5.48 million km^2 (38% of the degraded terrain). Loss of topsoil through water erosion is the most common type of soil degradation. It occurs in almost every country, under a great variety of climatic and physical conditions and land use. As the topsoil is normally rich in nutrients, a relatively large amount of nutrients is lost together with the topsoil. This process may lead to an impoverishment of the soil. Loss of topsoil itself is often preceded by compaction and/or crusting, causing a decrease in water infiltration capacity of the soil, and leading to an accelerated run-off and soil erosion. Loss of topsoil can also result from wind action. It is a widespread phenomenon in arid and semi-arid climates, but it also occurs under more humid conditions.

In general, coarse-textured soils are more susceptible to wind erosion than fine-textured soils. Wind erosion is nearly always caused by a decrease of the vegetation cover of the soil, due to overgrazing, pollution, salt stress, or removal of vegetation for domestic or for agricultural uses. In semi-arid climates natural wind erosion is often difficult to distinguish from human-induced wind erosion, but natural wind erosion is often aggravated by human activities. Growing conventional crops that require high inputs on these lands could increase rates of erosion and polluted runoff. Erosion is often associated with a reduction of vegetative cover, which increases the eroding power of the wind and water, which if not stopped in time ends up with desertification. Loss of nutrients and/or organic matter occurs if agriculture is practiced on poor or moderately fertile soils without sufficient application of manure or fertilizer. It causes a general depletion of the soils and leads to the decreases of plant biomass and land productivity. Loss of nutrients is a widespread phenomenon in countries where low-input agriculture is practiced. The rapid loss of organic matter after clearing the natural vegetation is also included in this type of soil degradation. The loss of nutrients by erosion of fertile topsoil is considered to be a side effect of erosion, and not distinguished separately.

The use of bioenergy crops on these lands, which are improper or non-sustainable for food production generate multiple ecological and economical benefits. For instance, these bioenergy crops could be used to stop land degradation and to restore soil fertility, thus it would act positively on carbon sequestration and reduce water pollution generated by the soil erosion. It would create biomass designated for bioenergy production, which will have a positive impact on the reduction of fossil-fuel consumption and it should generate financial incomes. Finally, the use of these degraded lands would not compete with lands designated for food production.

2.5.1 Abandoned Lands

In general, food crops are very nutrient demanding and are very sensitive to various stresses, thus they require active management that is very costly. Consequently, lands that are poorly productive and not economically sustainable (financial input > financial output) get abandoned. Most of the potential bioenergy crops can be grown on marginal lands, are less nutrient and water demanding than food crops and they do not require active care, thus they require a much lower financial input than food crops, suggesting that abandoned lands have the potential to be used for bioenergy crop production. Recently, Campbell *et al.*[42] estimated that the global area of abandoned agricultural land ranges between 3.85 million and 4.72 million km^2, which is 7.8–9.6% of the total agricultural lands (crop and pasture), and which, on average, produce about 4.3 t/ha/y of biomass. The replacement of the growing biomass by high yielding bioenergy crops has the added potential of restoring economic value to these lands.

Studies on WSGs such as switchgrass show the important role of WSGs in improving soil properties and controlling erosion.[43] Tall WSGs consisting of tall grass species produce abundant above and below ground biomass, with extensive deep root systems that can improve soil physical properties (porosity, fluxes of water and air), soil chemical, and biological properties (organic carbon and water contents, microbial processes,), and ultimately maintain soil productivity. Many species of *Salix*, *Populus*, and *Miscanthus* have characteristics of 'pioneer' species that show adaptations for growth on poor sites and under harsh conditions. Available data indicate that herbaceous and woody plants can improve soil characteristics, reduce soil water and wind erosion, and sequester soil organic carbon. Because of their deep root systems, warm season grasses also promote long-term carbon sequestration in deeper soil profile unlike row crops.

Growing dedicated energy crops in marginal and abandoned lands instead of fertile lands used for food crops will further benefit the soil and environment. Warm season grasses can grow in nutrient-depleted, compacted, poorly drained, acid, and eroded soils, thus representing good candidates for reclamation of marginal lands.[30] For example, WSGs can grow and persist in adverse conditions including compact, poorly drained, acid, and relatively contaminated soils. Varvel *et al.*[44] showed that predicted ethanol yield from switchgrass grown in marginal soil was greater than that from corn stover under the same fertilization conditions, showing that dedicated energy crops can be a viable option for producing renewable energy on these lands. Plantations of SRWCs can also be used to restore degraded soil. On sandy and clayey soils, conversion of crop land to aspen (*Populus deltoides*) plantations improved soil water retention.[45] Soils planted with SRWCs retain more soil water than those under cultivation and SRWCs have much greater cumulative water infiltration than row crops and pasture.[33] It was actually shown that soil erosion loss from row crops areas was about ten times higher than that in areas planted with SRWCs.[46]

Some abandoned lands are created because of land degradation and erosion, which caused the lost of productivity, and are still subject of further degradation. Both WSGs and SRWCs can also control wind erosion. Switchgrass grows a tall rigid stem and has deep rooting systems that confer resistance to erosive forces of wind. Switchgrass was shown to be an effective barrier against wind in semiarid regions,[47] and as a drought-tolerant species it can grow well in sandy and relatively windy environments. WSGs can be grown to control wind erosion near the soil surface,[48] especially in semiarid regions where wind erosion is more damaging than water erosion for soils. As a conclusion, WSGs have the capacity to reduce water and wind erosion, whereas perennial WSGs provide a permanent defense against wind erosion over croplands.

2.5.2 Dry Lands

Deserts are not considered as biologically productive land since they are considered that they do not contribute to the human sustainability (FAO definition). The expansion of bioenergy crop production on desert area would have the benefit that it would not compete with land that could be used directly for food production. Deserts are also constantly expanding, thus development of tolerant perennial plantation that could prevent or reduce soil erosion by fixing the ground would have also a very positive impact. It is estimated at 2.6 million km^2 of lands that need irrigation to keep their production potential, which in a near future will become subject to desertification if nothing is done. Very often excessive irrigation generates salinity problems and water erosion that in longer term will reduce the vegetative cover that will amplify the erosion. It is estimated that an additional 14 million km^2 of lands is suffering from low desertification, which consequently become improper for food production if nothing is done to stop it.[10] The following discussion will be focusing on strategies that could use part of the desert to produce bioenergy biomass and on advantages of the development of new bioenergy crop lands to stop desertification and to restore degraded lands.

All dry lands cannot, unfortunately, be used for biomass production; however, hedges and lands in desertification process (lands that no longer support irrigation) could be primary targets. In warm and dry areas, there is a large competition between irrigation and drinkable water, which sometime creates water over drafting. In addition, it is commonly observed that over-irrigation in these hot areas increases salinity issues. Both, consequently, render the soil unsuitable for major food crop cultivation (productivity is too low) and stimulate erosion. The management of these areas by the plantation of dry tolerant species, has the potential of first reducing water consumption, of stopping the extension of desertification, possible land restoration, and to block sandy winds. Since these land types became non-sustainable for food production and therefore are, most of the time, abandoned, they are a potential target for bioenergy crop production.

Desertification is a huge worldwide problem and it is constantly expanding. It is not affecting only food production; it is also not providing carbon sequestration since biological activity is minimal. The main reason of desertification is caused by a loss of vegetation due to overgrazing, repetitive drought stresses or salinity, which consequently leaves the land susceptible to erosion. The reintroduction of plant species tolerant to drought and salt stresses and that are not or are poorly subject to grazing can stop the erosion, thus halting desertification. Also, by stopping erosion, especially wind erosion, which is responsible of sandstorms commonly observed in China, Australia, and Africa, it will improve air and water quality. The second advantage is that restoring plant growth on 'non-productive land', will increase carbon sequestration underground since plants will have to develop a large rooting system often associated with a microbial community.[49,50] The third positive aspect is that this vegetation will also regenerate a new litter fall, which will reintroduce organic matter in the ground, and which after few years will restore soil fertility,[51] increase water retention, and, in a long-term perspective, might restore lands for food production. In addition, the production of valuable biomass on lands which were sterile (or almost), will generate a new economical activity by creating new incomes for farmers and could potentially give energy independence to isolated areas. Therefore, it has the potential to improve living conditions of these poor areas.

The presence of cities located on desert borders should favor the sustainability of biomass production for bioenergy since this biomass conversion to energy could be integrated with city waste conversion (see the waste section).

The proximity should integrate the use of sewage water from the cities to generate a short-term irrigation strategy in order to help the establishment of perennial biofuel crop plantations.[52–54] A general observation is that nutrient deficiency is accruing in many arid areas and eroded lands mainly as nitrogen or phosphorus deficiency in North America, North Africa, and Australia.[55] These nutrient deficiencies may cause desertification but may also be caused by desertification. Any stresses reducing plant biomass productivity and perennity will generate losses of organic matter in the soil (reduction of supply) and will induce soil erosion. These nitrogen- and phosphorus-deficient lands will affect strongly the restoration of vegetation.[56] Thus, the plant selection will have a key role in the re-establishment of new vegetative covers dedicated for biofuels. Legume species have the capability to fix atmospheric nitrogen to supply to their own needs and to enrich their environment with nitrogen. Mycorrhiza extends the plant's ability to absorb water and nutrients, in particular phosphorus.

The integration of a temporal irrigation system with sewage water should help to establish new vegetative covers and could prevent plant losses during extended drought periods.[53,54] Sewage water is known to be rich in nutrients and is often directly released into rivers, lakes, or oceans, creating eutrophication.[57] The integration of temporal irrigation systems with sewage water would also have the advantage to clean up this used water. These bioenergy crop lands would create a kind of natural filter where plants would absorb most of nutrients before the water reaches underground water reserves and rivers. For food crop irrigation, this water is not very suitable[58] since it can potentially carry some pathogens[59] and various heavy metals that accumulate in various plant organs: fruits, seeds, and leaves.[60] The use of sewage water on bioenergy crop lands would stay at a very low level per plant since theses plants would have been selected to grow in dry areas. Therefore, the excess of water could be used to extend the amount of irrigated surfaces and the bioenergy crop land areas. However, the biomass that will have been irrigated with sewage water will have to be processed with caution to avoid any risk of pathogen proliferation, (i.e. in fermenters or gasifiers) and should therefore be preferentially thermo-converted into bioenergy instead of being bio-converted into biofuels.

To increase the potential success of restoring biomass-producing covers on arid and semi-arid areas to reverse desertification, plant selection will be very important and the use of plants already adapted to these extreme conditions should be the first target. These plants should be able to resist drought and salt stresses, cope with grazing, have low nutrient requirements, strong biofuel properties such as a good oil, latex, or sugar content, good above ground productivity, and low ash content. In the longer term these species or new species could be further genetically engineered to improve their bioenergy conversion efficiency and their stability under various desert stresses. Perennial species are the most suitable plants for this project since their rooting stems will help to fix the topsoil layer against erosion and they should not require too much labor. The invasiveness of the species will most likely be insignificant since these plants will be growing under extreme conditions. There are only very few studies focusing on the selection of plant species that could be used to stop the extension of the desert. These studies were poorly developed probably due to under funding and due to the absence of real economical driving forces. The development of desert-adapted energy crops and the demonstration of the advantages and its sustainability should stimulate this research due to the economical potential and the great potential impacts on human sustainability.

Mineral composition and pH will be also determinant factors for the plant selection since they will significantly affect plant growth and land restoration.[61] A few studies showed that the supply of mycorrhizae spores or rhizobium bacteria helped the establishment of perennial species.[62] In similar lines, the use of legume trees (such as smoke tree and mesquite) in nitrogen poor lands like the Sonoran desert (USA) could be used in a desert restoration program.[55] There are also a few studies on desert-adapted plants for their potential as bioenergy crops. The crassulacean acid metabolism (CAM) type plants, such as *Agave* and *Opuntia ficus-indica* are well adapted to semi-arid conditions and can produce large amounts of biomass up to 43t DM/ha/y.[63] In addition CAM plants such as *Agave* plants contain a large amount of sugar that can be directly used for fermentation.[64] From the Euphorbiaceae species, Jatropha plants are very well adapted to various stresses and can grow in semi-arid regions. They produce seeds that contain between 27 and 40% of oils, which rends it very attractive for biodiesel production.[65–67] Some trees such as *Moringa tinctoria* and *Acacia seyal* are know for their wood fuels value and are able to grow in arid and semiarid areas and are already grown in desert.[68,69] In China, sand willows are grown in desert areas to stop desert expansion and it also became a biomass source for bioenergy production.[70,71] There are several other perennial species from semi-arid areas that have been evaluated for their bioenergy potential such as biomass yield and oil content[71] suggesting that plant diversity should be available to avoid bioenergy crop monoculture and to target various arid and semi-arid areas.

In summary, research is progressing to identify several plant species that would have strong biofuel potential and that could be used to stop land erosion and desertification. Diversity of plant species has an important role in the sustainability of plant-based bioenergy production. In addition, soils properties and composition differ between deserts and within an area. It will require adapted species to increase the success of plant restoration in semi-arid and arid areas. The advantages of bringing biofuel crops into arid and semi-arid areas, is that they will not compete with the arable lands used for food production and, in the long term, there is the potential to restore some lands for food production in poor areas. Unfortunately, there is not enough information yet to perform a complete evaluation of the impact of arid and semi-arid land restoration. The potential to use desert plant grown biomass as a bioenergy resource still has to be further analyzed to estimate how much marginal arable land could be saved for food crops or diversity preservation. Finally, if part of the desert would become suitable to produce some bioenergy crops, this area will probably have to be reconsidered as biologically productive land since it would contribute to human sustainability.

2.5.3 Land Polluted with Heavy Metals and Other Contaminants

The rapid industrial development that occurred worldwide in recent years has raised land pollution and environmental issues. Elevated concentrations of heavy metals in soils represent potential long-term environmental and health concerns because of their persistence in the environment and their associated toxicity to biological organisms. Furthermore, the costs of soil remediation also represent financial issues to landowners since costs only, and not income, are associated with land restoration. Agricultural land contamination by arsenic mainly originated from mineral extraction and waste processes, which are caused by poultry and swine feed additives, pesticides, and highly soluble arsenic trioxide stockpiles.[72] The consequences of heavy metal accumulation in the soil are not only associated with the storage of toxic elements in plant organs growing on these polluted land, but also to ground water and river contamination due to leaching, and have the risk to render the land unable to support any plant growth.[73] For example, an estimated 36 million people in the Bengal Delta are at risk from drinking arsenic-contaminated water.[74] It is estimated that soils affected by pollution cover an area of 0.22 million km^2 worldwide, of which 0.09 million km^2 is located in Europe.

Heavy metals (including arsenic, cadmium, chromium, copper, lead, mercury, nickel, and zinc) cannot be broken down into less harmful by-products, so phytoremediation strategies have been developed. They consist of using plants that can accumulate heavy metal in aboveground plant parts to remove them from the environment or to render them harmless.

High yielding biomass crops offer good potential for the phytoremediation of sites contaminated with heavy metals since it is not recommended to grow food crops on this land. This has the potential of creating financial income to restore polluted lands. Warm-season grasses and SRWCs can tolerate contaminated soils and be used as part of phytoremediation strategies.[75] Different biomass crops, species, and genotypes may show large differences in efficiency of heavy metal uptake, and in the concentration of metals in different plant parts. For phytoremediation purposes, it is desirable for metals to be concentrated in the harvested parts of the plant such as the stems or leaves of short rotation crops. For example the accumulation of pollutants in the leaves of SRWCs would allow a separation of leaf and stem organs and a processing of both organs into bioenergy independently. This separation would reduce the potential side effects of the pollutants in some conversion processes (bioconversion especially). Trees are now considered in phytoremediation strategies for heavy metal-contaminated lands.[76] In particular, willow (*Salix* spp.) encompasses characteristics required for both remediating and energy crops. Willow have been shown to take up large amounts of Cd and Zn. It can be propagated vegetatively and frequently harvested by coppicing, yielding as much as 10–15 t DM/ha/year. Bushy *Salix* species with erect stems, rapid growth, and good rooting ability are the most suitable for biomass coppice. In addition to high biomass productivity, *Salix* trees also have an effective nutrient uptake capacity and a pronounced capacity for heavy metal uptake, which allows them to colonize contaminated soils.[76] Recent studies also showed the ability for some poplar species to accumulate boron and selenium extracted from the soil,[77,78] as well as the potential for maize to remove zinc from moderately contaminated soils.[79] The contaminated nonfood biomass has a potential as a renewable energy source.[80,81] Thus, the generation of corn plants via genetic engineering that could be used for phytoremediation and that would not accumulate the pollutants into the seeds would have triple function: food, biofuels, and phytoremediation. Recent studies indicate that *Miscanthus* crops could be successfully grown on contaminated land, although high levels of heavy metals may reduce crop productivity. Most heavy metals accumulate in the roots and rhizomes, rather than in the harvested aerial parts.[82] In such a case, it would not be the ideal plant for phytoremediation since rhizomes removal would be required, but it still gives the option of growing bioenergy crops on contaminated lands.

Organic pollutants can often be converted by plants into less harmful metabolites. Research on hybrid poplars has demonstrated their ability to take up and effectively degrade or deactivate a number of contaminants, including atrazine, 1,4-dioxane, trinitrotoluene, and trichloroethylene.[83] Similarly, switchgrass tolerates soil contaminated with trinitrotoluene better than cool-season grasses such as tall fescue, and is effective at remediating soils contaminated with trinitrotoluene, atrazine, and metolachlor.[84–86]

Increasing plant tolerance and metabolism of organic chemicals or tolerance to heavy metals can be achieved using biotechnology. For example, plants can be engineered to absorb and metabolize higher amounts of metals and other pollutants by over-expressing modifying enzymes and transporters.[87,88] Understanding plant–microbe associations could also improve the efficiency of phytoremediation as many bacteria show a natural capacity to cope with contaminants.[89] A single genetically modified energy crop might be produced to efficiently take up several different pollutants, which would increase the effectiveness of phytoremediation of organic compounds and metals from contaminated sites, without impacting its biomass yield. It would offer the ability of using polluted lands, unsuitable for food crop production, to grow bioenergy crops with an economical value and to reduce environmental pollution via phytoremediation.

2.5.4 Saline Lands

Saline lands are estimated to represent 0.76 million km^2 worldwide, of which 0.53 million km^2 are present in Asia. Human-induced salinization can be the result of using irrigation water with 'high salt' content and mainly occurs under (semi-)arid conditions.[90] Salinization will also occur in coastal regions where seawater or fossil saline ground water intrudes on the ground water reserves of good quality used for irrigation. Human activities leading to an increase in evaporation of soil moisture in salt-containing soils can also induce salinization. The salinization of irrigated lands is increasingly detrimental to plant biomass production and agricultural productivity,[90,91] as most plant species are sensitive to high concentrations of sodium, which causes combined sodium toxicity and osmotic stress.

Thus salinity represents a significant land degradation and agricultural issue. For example, more than 50 % of the cropped land in Australia is affected by soil acidity, sodicity and salinity problems with an estimated annual impact to the agriculture of A\$2,559 million.[92] No food-crops can be grown sustainably on this type of soil, which however represent available lands for halophyte plants, which can tolerate high salt levels. In particular, *Arundo donax* is a crop that produce high biomass yield (45 t DM/ha/y) on saline lands. Furthermore, this perennial rhizomatous grass does not produce pollen and does not have fertile seeds, making it a good candidate bioenergy crop.[93,94] More conventional bioenergy crops such as certain poplar clones were shown to produce biomass when irrigated with landfill leachate enriched in sodium and chloride.[95]

In addition, identifying genes implicated in salt stress response as well as sodium exporters offers strategies for improving salt resistance in plants using biotechnology. Genomic regions that influence salt tolerance were identified in monocot crop plants such as wheat and rice, and they correspond to transporters that mediate the intracellular concentrations of sodium and potassium.[96]

The development of plants tolerating high levels of salt could allow biomass production on saline lands – estimated at 200 000 to 300 000 acres in California – and could potentially be used for the recycling of sewage water for agricultural irrigation.[37]

2.6 Waste Biomass

Several million km^2 of lands are available today to produce dedicated biomass for bioenergy production. From the actual cultivated lands, several billion tonnes of organic matter/biomass are produced and consumed as food or material, but part of it, residues, is considered as waste since it cannot be used for food or for material production. Part of this waste is efficiently used to improve farming land quality or is recycled as valuable products. However, most of these organic residues are accumulated in landfills to produce CO_2 and pollute the water or are applied in excess on farming lands, which make them also useless and produce CO_2 and pollutants. The use of these residues can be considered as biomass produced on free land since the use of this biomass for bioenergy production will not compete with the food consumption. Their utilization has also the potential to be beneficial for the environment since it can reduce pollution. These wastes could be divided in several categories depending of their origin, whether they have been transformed or not, and if they contain undesirable chemicals or not. To simplify the evaluation, the residues will be presented according to their land origin: forestland residues, farmland residues, and city land residues.

2.6.1 Forest Land Residues

Residues generated from forestry have two origins: (i) forest management consisting of tree residues produced during the growth of the forest and (ii) from wood residues generated during wood harvest and processing. Tree residues correspond to leaves, needles, branches, and bark that falls on the ground during the forest growth. Wood harvest and processing generates branches, bark, sawdust, and pulping wastes. These timber wastes represent 68–76% of the total above ground biomass.[97–99] Only a small proportion of timber waste is utilized to produce energy, and it is estimated that approximately 50% of the forest biomass will be left in the forest for degradation[100] and the rest (18–26%) is generated from offsite processing. The total wood (round wood) consumption in 2008 was estimated at 3448.6 million m^3 and 45% of it (1556.7 million m^3) was designated for industrial use, the rest, the wood fuels, representing 1892.0 million m^3, was used as for energy production.[101,145] In 1998 it was estimated by FAO that approximately 56% of the industrial wood was used for construction, 24% for paper and paperboard, and 20% as processed wood.[102] Between bark, sawdust, and logging residues the density varies between 320 and 400 kg/m^3 with 50–55% moisture content.[103] With the estimation of 60% of timber waste and 0.350 tonnes/m^3 and 50% moisture, the available biomass left for degradation corresponds to approximately 905.25 million tonnes of DM every year ([3448.6/0.40]× [60%] × [0.35] × [50%]; [timber + timber wastes] × [percentage of timber wastes] × [density of timber wastes] × [moisture content]). It was estimated that 500 million tonnes of wood is yearly consumed during the last 10 years for paper krafting and pulping,[145] to produce approximately 225 million tonnes of pulp (pulping yield estimated at 45%)[104] and therefore 1575 million tonnes of waste called 'black liquor'.[105] This black liquor can be burned on site to produce energy and to recover chemicals and the energy excess sold out. Techno-economic model analysis suggests that a combination of techniques including black liquor gasification and 'black liquor gasification–combined cycle' have the potential to produce more energy than is consumed by the pulping industry.[106,107] These improvements would reclassify the fourth largest industrial energy consumer, the pulp and paper industry,[18] to that of a net energy producer.

Leaf litter would be an additional option to generate biomass for bioenergy production if well managed. Only a certain percentage should be harvested to avoid organic matter depletion and associated consequences such as nutrient depletion, erosion, and desertification. The approximate litter yields from leaves and barks generated in deciduous forest is estimated at 300 tonnes/km^2/year for forest aged between 20 and 90 years old in a cool temperate climate.[108–110] Interestingly, the difference of litter production between 'evergreen gymnosperm' and deciduous angiosperm is very little, but the main difference is that deciduous forest generate one main leaf fall a year, in contrast to a whole year 'leaf fall' in the gymnosperm forest.[108]

To reduce harvesting time and harvesting cost, managed deciduous forest should be the main target since tree distribution on the ground can be organized. Part of these managed forests are used to produce timbers designated for paper production, construction wood, or bioenergy, and according to the Forest Resource Assessment (FRA) 2005,[111] it represents 3% (1.18 million km^2) all forest lands. The use of leaf litter from these forests has the potential to generate approximately 70.8 billion tonnes/year with a 5-year harvesting cycle. Due to the absence of study on leaf litter removal an arbitrary 5-year harvest cycle has been selected. A 5-year cycle should have a low impact on the organic matter accumulation and it should be more cost-effective to reduce the number of harvests than a small yearly harvest. An additional advantage of harvesting leaf litter, bark, and branches, is that it will reduce the accumulation of potential fire hazard material, therefore it should reduce the propagation of wildfires.

In summary, the conversion of the entire wood industry waste has the potential to cover almost 11.4% (44.673 × 10^{18} J/year) of the yearly worldwide energetic consumption of fossil fuels; 19.915 × 10^{18} J/year from industrial wood residues (905.25 billion kg/year × 22 MJ/kg),[99] 23.625 × 10^{18} J/year from black liquor (1575 billion kg/year × 15 MJ)[112] and 1.133 × 10^{18} /year from litter falls (70.8 billion kg/year × 16 MJ).[99]

2.6.2 Farmland Residues

There are two main types of wastes generated on farmlands, straw residues and manures. Straw residues, stovers, are all generated from seed or sugar production, thus they will vary between countries, according to their main agricultural production. According to a study by Kim and Dale,[113] based on cropping surface and average yield between 1997 and 2001, 1387.7 million tonnes of straw and 180 million tonnes of bagasse would be available for bioenergy conversion. It corresponds to 751 million tonnes from rice straw, 354 million tonnes from wheat straw, 203.6 million tonnes from corn stover, 58 million tonnes from barley, 10.8 million tonnes of oat, 10.3 million tonnes of sorghum, and 180 million tonnes of sugarcane bagasse. In their study, they calculated that 60% land cover with straw residues are required to maintain the level of soil organic manure, and thus maintain the land quality.[114] The energy value stored in dry grass straw, corn stover, and bagasse is estimated to vary between 15.4 and 19.4[115] because of the variability in ash and carbon contents. The conversion into bioenergy of the entire grass-derived residues, minus the left over residue requirement for soil fertility maintenance, has the potential to feed 7% (27.3 × 10^{18} J/year) of the yearly worldwide energetic consumption derived from fossil fuels, using 17.4 MJ/kg as the energy content average per kg of dry biomass (1567.7 million tonnes × 17.4 MJ/kg).[115]

The second type of waste generated in large quantities from farming is manure. This is generated by intensive livestock production, mainly beef, swine, diary, and poultry to produce meat, milk, and eggs, designated for human nutrition. Part of it is efficiently used as fertilizer on crop fields, and the leftover is often applied in excess as fertilizer and therefore creates land, water, and air pollution. The land application is usually seasonal since part of the time the land is covered with crops, or the land is free but it is at the end of the growing season, and therefore the application would generate a lot of nutrient leaching during fall and winter time, thus causing water pollution.[116,117] One of the consequences is that manure needs to be stored between land utilization cycles, which also creates some pollution when effluents are not collected and due to CO_2 losses and the production of greenhouse gases caused by an anaerobic fermentation.

The energy value of manure produced by intensive farming was estimated at 1700 million tonnes/year in 1999, which represents approximately a total energy of 25.5×10^{18} J/year (1700 million tonnes/year $\times$ 15.5 GJ/tonne)[118] and which is equivalent to 6.5% of the yearly worldwide energetic consumption derived from fossil fuels. This percentage has to be interpreted with caution since it corresponds to 100% conversion of the manure into energy. In practice to keep the 'biofertilizer' value of the converted manure, only a part of the manure will be converted into energy. One of the common points between manure and the other farming residues is the organic content. In contrast the main differences are the water content which is very high in manure (except poultry manure) and the mineral composition (N, P, K, and many others) is very high, which render them not very suitable for thermo conversion or to conventional enzymatic hydrolysis (external supply of hydrolytic enzymes). The most efficient conversion system seems to be anaerobic digestion, which produces energy and biogas and reduces the amount of biomass and keeps an high nutrient availability in the residue, which can still, further on, be used as fertilizer and be applied on the fields.[119] There are several case studies that demonstrated the feasibility of manure conversion into biogas.[119–121] Some studies are also completed with mixed wastes consisting of mixing plant residues with the manures[122,123] or mixed with industrial organic residues[124] to improve the conversion efficiency into biogas.

2.6.3 Urban Land Residues

The surface covered by cities and diverse constructions represents approximately 3.5 million km^2 of land[125] in which city area is estimated around 0.75 million km^2.[126] In urban lands, the primary productivity is almost insignificant since most of it is covered by concrete, asphalt, and few personal gardens and small parks; however, these areas were able to generate more than 2.02 billion tonnes of solid waste worldwide in 2006, according to the Global Waste Management Market Assessment from 2007.[127] These residues are known as municipal/urban solid waste (MSW) and are largely composed of organic wastes and are mainly exported to and accumulated in landfills, as only a little fraction is recycled. The organic fraction of MSW is primarily composed of paper, cardboard wastes, green wastes (from garden and landscaping), and food waste derivatives. This biodegradable component represents approx. 50–70% of the MSW (i.e. 53% in USA, 48.9% in Spain, 46–64% in Asia).[128–130] One of the biggest issues with landfills is the generation of pollutants, and the organic fraction is one of the main contributors. This biomass is rich in minerals and heavy metals that are slowly released by microbes and leached by water to end up in the rivers and ground water when they are not drained up by specific collectors.[131,132] In addition, this biomass also contributes significantly to the production of greenhouse gas effects; it releases mostly methane and CO_2.[132–134] These gases are mainly generated from the lignocellulosic biomass (paper, cardboard wastes, cooking oils, and green wastes) composed of cellulose, hemicellulose, and lignin. As a result, the recycling of the lignocellulosic biomass would provide a great source material for bioenergy production, which would consequently reduce the emission of greenhouse gases and the amount of MSW, which is accumulated in landfills. Of course the recycling of plastics, and other recyclable materials, would also contribute significantly to the reduction of landfill pollution and could be used for energy production.

Several studies are focusing on the transformation of the MSW organic biomass into energy. Recently, Alle Zihao Shi and co-workers[135] reported that 82.9 billion liters of waste paper-based biofuels could be produced in the world. Wasted cooking oils consist of vegetable oils (corn peanut, sunflower, olive, and soybean), which after processing could be used as biodiesel.[136] Several methods have been already developed such as alkali-catalyzed transesterification, hydroprocessing, and enzymatic conversion to produce free fatty acids and fatty acid methyl esters.[137–139] Finally, there are also lots of food wastes derived from specific industrial processes such as grape and tomato skins and seeds.[140,141] Since the composition of this type of organic waste is well defined, specific bio- and thermo-conversion systems could be established to produce bioenergy. For example, 5 million tonnes of citrus peel are produced in Florida every year. Verma and co-workers[142] demonstrated that this biomass could be reused and transformed into ethanol after enzymatic hydrolysis and fermentation. In summary, the conversion of the entire biomass of MSW would generate 14×10^{18} J/year, and the incorporation of the non-biogenic fraction (i.e. plastics and rubber) would add an additional 11×10^{18} J/year, which represent 3.6% and 2.3% respectively or 5.9% of the yearly worldwide energetic consumption of fossil fuels.[143]

2.7 Conclusions

Worldwide energy consumption was estimated at 498×10^{18} J in 2006, from which 393×10^{18} J are originated from fossil fuels.[18] The importance of generating energy from photosynthetic organisms is to close the carbon cycle loop and reduce or stop carbon loading of the cycle with fossil fuels. In 2006, to replace the energy derived from fossil fuels, it would require the use of at least 28 million km^2 of the highest quality of marginal lands to produce the equivalent energy from plant biomass. Therefore, selecting and designing plants well adapted to marginal lands and with improved efficiency to harvest light and to fix CO_2 could achieve the reduced land requirements. Also the increase of the calorific value of the biomass as well as the efficiency of its conversion would reduce required surfaces to generate enough energy. For example, the lignin has a much higher calorific value than cellulose and hemicellulose (21.2 MJ/tonnes DM and 17.5 MJ/tonnes DM, respectively)[144] and represents only 25% of the biomass. The enrichment in oil content of the biomass without affecting plant yield would also increase the energetic values and could be used almost directly as liquid fuel after extraction. In addition to the calorific value of the biomass, a bidirectional adaptation of dedicated energy crops and conversion approaches will reduce drastically the energetic cost of conversion processes.

Another important aspect to consider is the conversion of organic wastes into energy. These 'leftovers' are already available from forest and wood processing, farm residues and wastes, and urban organic residues, and would fulfill 22.4% ($44.7 + 52.8 + 14 \times 10^{18}$ J) of the total worldwide energy consumption (498×10^{18} J), which consequently would reduce land needs by 22.4%. These organic wastes are converted into CO_2 naturally by biological conversion processes. The difference here is that with the same amount of CO_2 released into the atmosphere, some energy will be harvested and could potentially feed 20% of human energy consumption.

Growing bioenergy crops on non-food lands gives more than renewable energy. Increasing biomass production on marginal lands, degraded lands, and desert will also increase the accumulation of organic matter in the soil since biomass accumulations aboveground and below ground of plant organs are positively correlated. Thus, growing bioenergy crops on non-cultivated lands will also participate significantly in long-term carbon sequestration. Finally, since growing bioenergy crops will generate economical outcome, easy accessible polluted and eroded lands will become suitable to grow perennial plants for biomass, thus they can be used to either stop further erosion or help with phytoremediation. In a longer-term perspective, developing production of bioenergy crops could potentially restore to these areas the quality required to grow food crops again.

Acknowledgements

This work conducted by the Joint BioEnergy Institute was supported by the Office of Science, Office of Biological and Environmental Research, of the US Department of Energy under Contract No. DE-AC02-05CH11231.

References

1. I. N. Forseth, *in Nature Education Knowledge*, Nature Publishing Group, 2010, vol. 1.
2. *Terrestrial Global Productivity*, Academic Press, London, 2001.
3. T. C. Whitmore, *An Introduction to Tropical Rain Forests*, 2nd edn, Oxford University Press, Oxford, 1998.
4. D. Burslem, N. Garwood and S. Thomas, *Science*, 2001, **291**, 606.

5. J. Bruinsma, *World Agriculture: Towards 2015/2030: An FAO Perspective*, Earthscan Publications Ltd, London, 2003.
6. G. Fischer, S. Prieler and H. van Velthuizen, *Biomass and Bioenergy*, 2005, **28**, 119.
7. G. Daily, *Science*, 1995, **269**, 350.
8. K. Weibe and E. R. S. Resource Economics Division, *United States Department of Agriculture*, Economic Research Service, Washington, DC, Agricultural Economics Reports,2003.
9. P. L. G. Vlek, D. Hillel and A. K. Braimoh, *in Land Use and Soil Resources*, eds. A. K. Braimoh and P. L. G. Vlek, Springer Netherlands, 2008, pp. 1001–1119.
10. H. R. Eswaran, R. Lal and P. F. Reich, *Land Degradation: An Overview*, Oxford University Press, New Delhi, India, 2001.
11. H. Breman, *Biotropica*, 1992, **24**, 328.
12. J. Ibanez, J. Martinez and S. Schnabel, *Ecol. Model.*, 2007, **205**, 277.
13. H. Dregne, *Economic Geography*, 1977, **53**, 9.
14. F. Chapin and G. Shaver, *Ecology*, 1985, **66**, 564.
15. A. McGuire, J. Melillo, D. Kicklighter, Y. Pan, X. Xiao, J. Helfrich, B. Moore, C. Vorosmarty and A. Schloss, *Global Biogeochemical Cycles*, 1997, **11**, 173.
16. X. Xiao, D. Kicklighter, J. Melillo, A. McGuire, P. H. Stone and A. P. Sokolov, *Responses of Primary Production and Total Carbon Storage to Changes in Climate and Atmospheric CO_2 Concentration*, Massachusetts Institute of Technology, Cambridge, Massachusetts,1995.
17. R. Fuentes and C. Taliaferro, in *Trends in New Crops and New Uses*, eds. J. Janick and A. Whipkey, ASHS Press, Alexandria, VA, 2002, p. 599.
18. IEA-Report, *Key World Energy Statistics*, International Energy Agency, Paris, France, 2008.
19. R. C. Francis, R. B. Hanna, S.-J. Shin, A. F. Brown and D. E. Riemenschneider, *Biomass and Bioenergy*, 2006, **30**, 803.
20. B. Greaves, N. Borralho and C. Raymond, *Forest Science*, 1997, **43**, 465.
21. N. D. Kien, T. H. Quang, G. Jansson, C. Harwood, D. Clapham and S. von Arnold, *Annals of Forest Science*, 2009, **66**, 711.
22. M. Lindstrom, H. Jameel, V. Naithani, A. Kirkman and J. Renard, in *Materials, Chemicals, and Energy from Forest Biomass*, ed. D. S. Argyropoulos, Oxford University Press, 2007, p. 613.
23. C. Walter, S. Carson, M. Menzies, T. Richardson and M. Carson, *World Journal of Microbiology and Biotechnology*, 1998, **14**, 321.
24. J. S. Yuan, K. H. Tiller, H. Al-Ahmad, N. R. Stewart and C. N. Stewart, *Trends Plant Sci.*, 2008, **13**, 421.
25. X.-G. Zhu, S. P. Long and D. R. Ort, *Annu. Rev. Plant Biol.*, 2010, **61**, 235.
26. S. Gan and R. M. Amasino, *Science*, 1995, **270**, 1986.
27. R. K. M. Hay and J. Roy Porter, *The Physiology of Crop Yield*, 2nd edn, Wiley-Blackwell, 2006.
28. O. Calderini, T. Bovone, C. Scotti, F. Pupilli, E. Piano and S. Arcioni, *Plant Cell Rep.*, 2007, **26**, 611.

29. P. R. H. Robson, I. S. Donnison, K. Wang, B. Frame, S. E. Pegg, A. Thomas and H. Thomas, *Plant Biotechnol. J.*, 2004, **2**, 101.
30. H. Blanco-Canqui, *Agronomy J.*, 2010, **102**, 403.
31. M. Sanderson, R. Reed, S. McLaughlin, S. Wullschleger, B. Conger, D. Parrish, D. Wolf, C. Taliaferro, A. Hopkins, W. Ocumpaugh, M. Hussey, J. Read and C. Tischler, *Bioresour. Technol.*, 1996, **56**, 83.
32. R. Miller and B. A. Bender, in *Proceedings of the Short Rotation Crops International Conference*, eds. J. R. S. Zalensy, R. Mitchell and J. Richardson, USDA Forest Service, Northern Research Station, Bloomington, Minnesota, 2008, p. 36.
33. L. Bharati, K. Lee, T. Isenhart and R. Schultz, *Agroforestry Systems*, 2002, **56**, 249.
34. M. D. Coleman, J. G. Isebrands, D. N. Tolsted and V. R. Tolbert, *J. Environ Manag.*, 2004, **33**, S299.
35. A. H. Heggenstaller, R. P. Anex, M. Liebman, D. N. Sundberg and L. R. Gibson, *Agronomy J.*, 2008, **100**, 1740.
36. E. Malezieux, Y. Crozat, C. Dupraz, M. Laurans, D. Makowski, H. Ozier-Lafontaine, B. Rapidel, S. de Tourdonnet and M. Valantin-Morison, *Agronomy for Sustainable Development*, 2009, **29**, 43.
37. D. Putnam, *38th California Alfalfa & Forage Symposium and Western Alfalfa Seed Conference*, UC-Davis California, USA, 2008.
38. G. Pilate, E. Guiney, K. Holt, M. Petit-Conil, C. Lapierre, J.-C. Leplé, B. Pollet, I. Mila, E. A. Webster, H. G. Marstorp, D. W. Hopkins, L. Jouanin, W. Boerjan, W. Schuch, D. Cornu and C. Halpin, *Nature Biotechnol.*, 2002, **20**, 607.
39. Y. Xi, Y. Ge and Z. Wang, *Methods Mol. Biol.*, 2009, **581**, 53.
40. M. G. S. Fernandez, P. W. Becraft, Y. Yin and T. Luebberstedt, *Trends Plant Sci.*, 2009, **14**, 454.
41. L. R. Oldeman, R. T. A. Hakkeling and W. G. Sombroek, *World Map of the Status of Human-induced Soil Degradation: An Explanatory Note*, Global Assessment of Human-induced Soil Degradation (GLASOD), 1991.
42. J. E. Campbell, D. B. Lobell, R. C. Genova and C. B. Field, *Environ. Sci. Technol.*, 2008, **42**, 5791.
43. D. Kemper, S. Dabney, L. Kramer, D. Dominick and T. Keep, *J. Soil Water Conserv.*, 1992, **47**, 284.
44. G. E. Varvel, G. E. Vogel, R. B. Mitchell, R. F. Follett and J. M. Kimble, *Biomass Bioenergy*, 2008, **32**, 18.
45. I. Messing, A. Alriksson and W. Johansson, *Soil Use and Management*, 1997, **13**, 209.
46. D. Pimentel and J. Krummel, *Biomass*, 1987, **14**, 15.
47. J. Bilbro and D. Fryrear, *J. Soil Water Conserv.*, 1997, **52**, 447.
48. H. Blanco-Canqui and R. Lal, Principles of Soil Conservation and Management, *SpringerLink*, 2008.
49. V. Saul-Tcherkas and Y. Steinberger, *Soil Biol. Soil Biochem.*, 2009, **41**, 1882.

50. A. Porras-Alfaro, J. Herrera, R. L. Sinsabaugh, K. J. Odenbach, T. Lowrey and D. O. Natvig, *Appl. Environ. Microb.*, 2008, **74**, 2805.
51. S. A. Ewing, R. J. Southard, J. L. Macalady, A. S. Hartshorn and M. J. Johnson, *Soil Sci. Soc. Am. J.*, 2007, **71**, 469.
52. R. F. Daubenmire and H. E. Charter, *Botanical Gazette*, 1942, **103**, 762.
53. Y. A. Saadat, M. M. Jahromi and A. M. Hassanli, *World Appl. Sci. J.*, 2009, **7**, 1239.
54. G. Singh and M. Bhati, *J. Plant Nutr.*, 2003, **26**, 2469.
55. R. A. Virginia, in *Environmental Restoration: Science and Strategies for Restoring the Earth*, ed. J. J. Berger, Island Press, Washington DC, 1st edn, 1990, p. 412.
56. M. Bowker, J. Belnap, D. Davidson and S. Phillips, *Ecol. Appl.*, 2005, **15**, 1941.
57. P. J. Oberholster, A. M. Botha and T. E. Cloete, *Environ. Pollution*, 2008, **156**, 184.
58. O. Amahmid, S. Asmama and K. Bouhoum, *Int. J. Food Microbiol.*, 1999, **49**, 19.
59. K. Ibenyassine, R. A. Mhand, Y. Karamoko, B. Anajjar, M. Chouibani and M. M. Ennaji, *J. Environ. Health*, 2007, **69**, 47.
60. M. I. Lone, T. Mahmood, K. Saifullah and G. Hussain, *Int. J. Agric Biol.*, 2003, **5**, 533.
61. F. Huerta-Martinez, J. Vazquez-Garcia, E. Garcia-Moya, L. Lopez-Mata and H. Vaquera-Huerta, *Plant Ecol.*, 2004, **174**, 79.
62. N. Requena, E. Perez-Solis, C. Azcon-Aguilar, P. Jeffries and J. Barea, *Appl. Environ. Microb.*, 2001, **67**, 495.
63. P. Nobel, *New Phytologist*, 1991, **119**, 183.
64. A. M. Borland, H. Griffiths, J. Hartwell and J. A. C. Smith, *J. Exp. Bot.*, 2009, **60**, 2879.
65. L. Canoira, R. Alcantara, J. Garcia-Martinez and J. Carrasco, *Biomass and Bioenergy*, 2006, **30**, 76.
66. W. M. J. Achten, L. Verchot, Y. J. Franken, E. Mathijs, V. P. Singh, R. Aerts and B. Muys, *Biomass and Bioenergy*, 2008, **32**, 1063.
67. W. Achten, E. Mathijs, L. Verchot, V. Singh and R. Aerts, Biofuels, *Bioproducts & Biorefining*, 2007, **1**, 283.
68. M. Seth, *Bot. Rev.*, 2003, **69**, 321.
69. E. S. Ariga, *J. Sustainable Agric.*, 1997, **10**, 25.
70. OLGPSDS-report, *Review Sustainable Development in China – Agriculture, Rural Development, Land, Drought and Desertification*, The Office of the Leading Group for Promoting the Sustainable Development Strategy, P. R. China (Department of Regional Economy National Development and Reform Commission), 2008.
71. S. Xiong, Q.-G. Zhang, D.-Y. Zhang and R. Olsson, *Bioresour. Technol.*, 2008, **99**, 479.
72. T. F. Jones, *Poultry Sci.*, 2007, **86**, 2.

73. R. V. Galiulin, V. N. Bashkin, R. R. Galiulina and P. Birch, *Land Contamination & Reclamation*, 2001, **9**, 349.
74. D. K. Nordstrom, *Science*, 2002, **296**, 2143.
75. D.-H. Kang, L. Y. Hong, A. P. Schwab and M. K. Banks, *J. Environ. Sci. Health*, 2008, **43**, 627.
76. I. Pulford and C. Watson, *Environ. Int.*, 2003, **29**, 529.
77. G. Bañuelos, M. Shannon, H. Ajwa, J. H. Draper, J. Jordahl and L. Licht, *Int. J. Phytoremediation*, 1999, **1**, 81.
78. B. H. Robinson, S. R. Green, B. Chancerel, T. M. Mills and B. E. Clothier, *Environ. Pollution*, 2007, **150**, 225.
79. E. Meers, S. Van Slycken, K. Adriaensen, A. Ruttens, J. Vangronsveld, G. Du Laing, N. Witters, T. Thewys and F. M. G. Tack, *Chemosphere*, 2010, **78**, 35.
80. M. Ghosh and S. Singh, *Appl. Ecol. Environ. Res.*, 2005, **3**, 1.
81. N. Witters, S. V. Slycken, A. Ruttens, K. Adriaensen, E. Meers, L. Meiresonne, F. M. G. Tack, T. Thewys, E. Laes and J. Vangronsveld, *BioEnergy Res.*, 2009, **2**, 144.
82. B. Ezaki, E. Nagao, Y. Yamamoto, S. Nakashima and T. Enomoto, *Plant Cell Rep.*, 2008, **27**, 951.
83. J. Burken and J. Schnoor, *Environ. Sci. Technol.*, 1998, **32**, 3379.
84. G. Krishnan, G. Horst, S. Darnell and W. Powers, *Environ. Pollution*, 2000, **107**, 109.
85. T. Chekol, L. Vough and R. Chaney, *Int. J. Phytoremediation*, 2002, **4**, 17.
86. F. Zhao, E. Lombi and S. McGrath, *Plant and Soil*, 2003, **249**, 37.
87. D. N. Dowling and S. L. Doty, *Curr. Opin. Biotechnol.*, 2009, **20**, 204.
88. A. R. Memon and P. Schroder, *Environ. Sci. Pollution Res.*, 2009, **16**, 162.
89. N. Weyens, D. van der Lelie, S. Taghavi, L. Newman and J. Vangronsveld, *Trends Biotechnol.*, 2009, **27**, 591.
90. A. Banin and A. Fish, *Environ. Monitoring and Assessment*, 1995, **37**, 17.
91. J. S. Rawat and S. P. Banerjee, *Plant and Soil*, 1998, **205**, 163.
92. C. Williams, T. Biswas, G. Schrale and J. Virtue, Irrigation Australia 2008 Conference, *Melbourne*, 2008.
93. I. Lewandowski, J. Scurlock, E. Lindvall and M. Christou, *Biomass and Bioenergy*, 2003, **25**, 335.
94. W. F. Anderson, B. S. Dien, S. K. Brandon and J. D. Peterson, *Appl. Biochem. Biotechnol.*, 2008, **145**, 13.
95. J. A. Zalesny, R. S. Zalesny, A. H. Wiese, B. Sexton and R. B. Hall, *Environ. Pollution*, 2008, **155**, 72.
96. T. Horie, F. Hauser and J. I. Schroeder, *Trends Plant Sci.*, 2009, **14**, 660.
97. T. Abebe and S. Holm, *Int. For. Rev.*, 2003, **5**, 45.
98. T. Tahvanainen, H. Kilpeläinen, L. Sikanen, H. Heräjärvi, J. Lindblad and E. Verkasalo, *Wood Material Sci. Eng.*, 2008, **3**, 1.
99. J. Akarakiri, *Energy*, 1993, **18**, 763.
100. A. Demirbas, *Energy Sources*, 2009, **31**, 1687.

101. Z. Wenjun, *Environ. Monitoring and Assessment*, 2007, **125**, 301.
102. E. Matthews, *From Forests to Floorboards: Trends in Industrial Round-wood Production and Consumption*, EarthTrends 2001 World Resources Institute, 2001.
103. F. Wikstrom, *Biomass and Bioenergy*, 2007, **31**, 40.
104. A. Terdwongworakul, V. Punsuwan, W. Thanapase and S. Tsuchikawa, *J. Wood Sci.*, 2005, **51**, 167.
105. C. Biermann, *Essentials of Pulping and Papermaking*, Elsevier Science & Technology, San Diego, CA, 1993.
106. H. Eriksson and S. Harvey, *Energy*, 2004, **29**, 581.
107. S. Farahani, E. Worrell and G. Bryntse, Resources, *Conservation and Recycling*, 2004, **42**, 317.
108. J. Bray and E. Gorham, *Adv. Ecol. Res.*, 1964, **2**, 101.
109. R. M. Hurd, *Ecology*, 1971, **52**, 881.
110. A. Raizada and M. Srivastava, *Forest Ecology and Management*, 1986, **15**, 215.
111. FRA-Report, *Global Forest Resources Assessment 2005: 15 Key Findings*, Food and Agriculture Organization of the United Nations, Rome, Italy, 2005.
112. A. Demirbas, *Energy Conversion and Management*, 2002, **43**, 877.
113. S. Kim and B. Dale, *Biomass and Bioenergy*, 2004, **26**, 361.
114. W. W. Wilhelm, J. M. E. Johnson, D. L. Karlen and D. T. Lightle, *Agronomy J.*, 2007, **99**, 1665.
115. T. Tew and R. Cobill, in *Genetic Improvement of Bioenergy Crops*, ed. W. Vermerris, SpringerLink, New York, NY, 2008, p. 450.
116. R. Nicholson, J. Webb and A. Moore, *Biosystems Eng.*, 2002, **81**, 363.
117. H. Steinfeld and T. Wassenaar, *Annu. Rev. Environ. Res.*, 2007, **32**, 271.
118. J. Sheffield and H. Zygmunt, ed.*Department of Energy*, 1999.
119. K. B. Cantrell, T. Ducey, K. S. Ro and P. G. Hunt, *Bioresour. Technol.*, 2008, **99**, 7941.
120. S. R. Harper, C. C. Ross and G. E. Valentine, *J. Water Pollution Control Federation*, 1988, **60**, 876.
121. G. J. Escobar and M. A. Heikkilä, *Biogas Production in Farms*, Through Anaerobic Digestion of Cattle and Pig Manure. Case Studies and Research Activities in Europe, TEKES – the Finnish Funding Agency for Technology and Innovation, Länsi-Pasila, 1999.
122. R. Summers, P. Hobson, C. Harries and A. Richardson, *Biol. Wastes*, 1987, **20**, 43.
123. J. Lu, L. Zhu, G. Hu and J. Wu, *Biomass and Bioenergy*, 2010, **34**, 821.
124. E. Comino, M. Rosso and V. Riggio, *Bioresour. Technol.*, 2009, **100**, 5072.
125. CIESIN, Center for International Earth Science Information Network (CIESIN), International Food Policy Research Institute (IFPRI); Centro Internacional de Agricultura Tropical (CIAT). 2004. Global Rural-Urban Mapping Project (GRUMP), Socioeconomic Data and Applications Center (SEDAC), 2004 edn, 2005.
126. A. Schneider, M. A. Friedl and D. Potere, *Environ. Res. Lett.* 2009, 4, 0044003.

127. http://www.researchandmarkets.com/reportinfo.asp?report_id=461875, Global Waste Management Market Assessment 2007.
128. M. Kayhanian, *Waste Management Res.*, 1995, **13**, 123.
129. A. Garcia, M. Esteban, M. Marquez and P. Ramos, *Waste Management*, 2005, **25**, 780.
130. D. Hoornweg and L. Thomas, *What a Waste: Solid Waste Management in Asia*, Washington, DC, 1999.
131. D. Fatta, A. Papadopoulos and M. Loizidou, *Environ. Geochem. Health*, 1999, **21**, 175.
132. N. Wilson, L. Peissig, P. Burford, A. Radians, R. Burnyeat and B. Sielaff, *Land Treatment of Landfill Leachate*, Minnesota Pollution Control Agency, 2007.
133. J. Bogner and E. Mathews, *Global Biogeochem. Cycles*, 2003, **17**, 31.
134. X. F. Lou and J. Nair, *Bioresour. Technol.*, 2009, **100**, 3792.
135. A. Z. Shi, L. P. Koh and H. T. W. Tan, *Global Change Biology – Bioenergy*, 2009, **1**, 317.
136. Y. Zhang, M. A. Dube, D. D. McLean and M. Kates, *Bioresour. Technol.*, 2003, **90**, 229.
137. Z. Helwani, M. R. Othman, N. Aziz, W. J. N. Fernando and J. Kim, *Fuel Processing Technol.*, 2009, **90**, 1502.
138. C. Komintarachat and S. Chuepeng, *Ind. Eng. Chem. Res.*, 2009, **48**, 9350.
139. S. Bezergianni, A. Dimitriadis, A. Kalogianni and P. A. Pilavachi, *Bioresour. Technol.*, 2010, **101**, 6651.
140. R. Xu, L. Ferrante, C. Briens and F. Berruti, *J. Anal. Appl. Pyrolysis*, 2009, **86**, 58.
141. V. Mangut, E. Sabio, J. Ganan, J. Gonzalez, A. Ramiro, C. Gonzalez, S. Roman and A. Al-Kassir, *Fuel Processing Technol.*, 2006, **87**, 109.
142. D. Verma, A. Kanagaraj, S. Jin, N. D. Singh, P. E. Kolattukudy and H. Daniell, *Plant Biotechnol. J.*, 2010, **8**, 332.
143. E. I. A. (EIA), in *Energy Information Administration/Methodology for Allocating MSW to Biogenic/Non-Biogenic Energy*, ed. U. S. Department of Energy, Energy Information Administration, Washington DC, 2007.
144. E. S. Domalski and T. A. Milne, *Thermodynamic Data for Biomass Materials and Waste Components*, The American Society of Mechanical Engineers, New York, 1987.
145. http://faostat.fao.org

CHAPTER 3

Surface Science Studies Relevant for Metal-catalyzed Biorefining Reactions

J. WILL MEDLIN

Associate Professor and ConocoPhillips Faculty Fellow, Colorado Center for Biorefining and Biofuels, Renewable and Sustainable Energy Institute, Department of Chemical and Biological Engineering, University of Colorado, Boulder, CO, 80309-0424

3.1   Introduction

The refining of biomass into fuels and chemicals presents important challenges for heterogeneous catalysis. As with petroleum refining, catalysts are needed to facilitate deconstruction and upgrading reactions of the raw material, and some reactions and catalysts that can be utilized in biorefineries are similar to reactions in petroleum refineries. However, biomass refining presents some unique challenges. Whereas petroleum feedstocks are largely composed of long-chain hydrocarbons, biomass feedstocks and deconstruction intermediates have a high content of oxygen; in many cases (as in carbohydrates), the oxygen content is similar to the carbon content. The high oxygen content creates both challenges and opportunities: highly oxygenated species are generally unsuitable as fuels, but oxygenated functional groups can serve as 'handles' that can be used to perform a variety of catalytic chemistries to produce value-added chemicals and fuels. The importance of catalyzing reactions of highly oxygenated compounds has sparked major new research activity in heterogeneous catalysis, as new

RSC Energy and Environment Series No. 4
Chemical and Biochemical Catalysis for Next Generation Biofuels
Edited by Blake Simmons

Published by the Royal Society of Chemistry, www.rsc.org

materials capable of actively and selectively catalyzing reactions of oxygenates are much sought after.

Another key difference in the refining of biomass is the low volatility of many biomass-derived reagents; this low volatility is in fact a consequence of the functionality, which causes strong interactions between molecules as well as with surfaces, resulting in significantly higher boiling temperatures. Many of these high-boiling compounds degrade or polymerize when one attempts to boil them to produce a vapor, making it more difficult to directly investigate the chemistry of highly functional oxygenates at a catalyst–vapor or catalyst–vacuum interface. As discussed below, many reagents in biorefining are introduced to reactions in water solvents, and the water has the potential to strongly affect surface chemistry. However, because many of the most desirable biofuels are hydrocarbons or lightly oxygenated compounds, the products of many key biorefining reactions are hydrophobic, so that biphasic reaction conditions are observed. The situation of biomass refining is therefore significantly different than typical petroleum refining conditions, which often occur in the vapor phase and very rarely occur with aqueous solvent.

The objective of this contribution is to review mechanistic surface science studies aimed at elucidating the unique features of heterogeneous catalysis in biorefining processes. The main objective of surface science studies is generally to develop a detailed understanding of surface reaction mechanisms using spectroscopic (and, increasingly, computational) techniques. The rationale for such investigations, which are typically conducted on model surfaces under very low pressures, is that detailed structure–property relations will enable design of improved catalysts. We begin by discussing recent contributions made to the understanding of more traditional, vapor-phase chemical reactions that do not involve high oxygen functionality in the reacting species. A few examples are provided to illustrate the utility of surface reaction mechanism development in the design of improved catalysts. We then discuss some of the key reaction processes involved in biorefining operations to frame more specific questions related to heterogeneous catalysis. This is followed by a description of some of the major tools used in surface science investigations as well as the advantages and limitations of the surface science approach. The two main sections of the review then describe the surface chemistry of functional groups important in biomass refining on metal surfaces, with the first section devoted to molecules containing a single functional group and the second section discussing molecules containing two or more functional groups. Finally, connections between surface science investigations and heterogeneous catalysis studies of several key reactions of biomass-derived oxygenates are discussed, and needs for future surface-level studies are described.

3.2 Surface Science Contributions to Catalyst Design

A number of textbooks and other sources provide an overview of petroleum refining and associated technologies. The focus of this section, instead, is on illustrating how fundamental investigations of solid catalysts have led to an improved understanding of the relevant surface chemistry and has ultimately resulted in design of improved catalysts. This previous work provides concrete motivation for the use of model studies and surface science techniques for development of improved catalysts. For brevity, two examples of catalyst design based on surface science investigations are included, but several other examples are available in the literature.[1–8]

3.2.1 Ethylene Epoxidation

The selective epoxidation of ethylene to ethylene oxide (EO) is one of the highest value-generating processes in the chemical industry.[9,10] The process has generally been conducted over supported Ag catalysts, which show good activity and, more uniquely, high selectivity to EO rather than combustion products. A number of approaches have been explored for enhancing the activity and selectivity of the reaction. For example, halogenated and alkali promoters are added to the Ag catalyst to increase the EO selectivity. The mechanism by which these promoters influence reactivity has been the subject of investigation over many years.[9,11–13] Barteau and co-workers have extensively investigated and identified the reaction pathways involved in olefin epoxidation and combustion over Ag surfaces.[14–18] These researchers proposed that the EO selectivity is controlled by competing reaction pathways from a common intermediate, termed an oxametallacycle (–O–C–C–) species. Several studies have shown that selectivity can be improved by adjusting the catalyst surface composition or structure to stabilize the transition state for the desired reaction pathway over the undesired pathway.[14,19] These results indicate that important advances in catalyst composition can be achieved even for reactions that have been studies for many decades.

3.2.2 Acetylene Hydrogenation

The selective hydrogenation of acetylene to ethylene is a key step upstream of ethylene polymerization processes. High selectivity for hydrogenation of acetylene to ethylene, rather than ethylene to ethane, is desired to avoid consumption of valuable polymerization feedstock. Pd-based catalysts, in particular bimetallic catalysts such as PdAg, have been found to show excellent selectivity for the desired reaction. The mechanism for the selective hydrogenation has been explored in great detail using both computational and experimental techniques.[20–25] Norskov and co-workers recently used computational techniques and mechanistic information developed through previous surface science experimentation to screen alternative catalyst compositions that could maintain high selectivity while reducing precious metal content, and predicted that a NiZn bimetallic surface as a low-cost, effective alternative.[26] Experimental screening of the bimetallic catalyst confirmed the computational predictions, and again reveals the ability of a fundamental, mechanistic approach to generate a radically new catalyst for a much-studied reaction.

3.3 Biorefining Routes: Key Intermediates and Transformations

Many pathways requiring heterogeneous catalysts have been proposed for the conversion of lignocellulosic biomass to fuels and chemicals. A broad overview of several, representative approaches is briefly discussed here and summarized in Figure 3.1. Rather than attempting to provide a comprehensive overview of these approaches, the focus is on the types of intermediates (in terms of molecular structure) involved in each approach. Production of chemicals from biomass generally requires multiple steps. The early steps in the process involve deconstruction of biomass into components such as synthesis gas, sugars, or other oxygenates. Later steps involve the upgrading of the fuel through reactions such as deoxygenation, coupling, oligomerization, etc. Some of the basic pathways are discussed in the following subsections.

3.3.1 Biomass Gasification Followed by Synthesis Gas Upgrading

This technology generally involves the use of high temperatures, steam, and oxygen to convert biomass to synthesis gas, i.e. CO and H_2, along with side products such as CO_2 and hydrocarbon 'tars'.[27,28] The synthesis gas stream can then be converted to hydrocarbon products using a number of routes, including Fischer-Tropsch synthesis and methanol synthesis followed by methanol-to-gas (MTG) catalysis.[28] A major advantage of gasification-based approaches is their

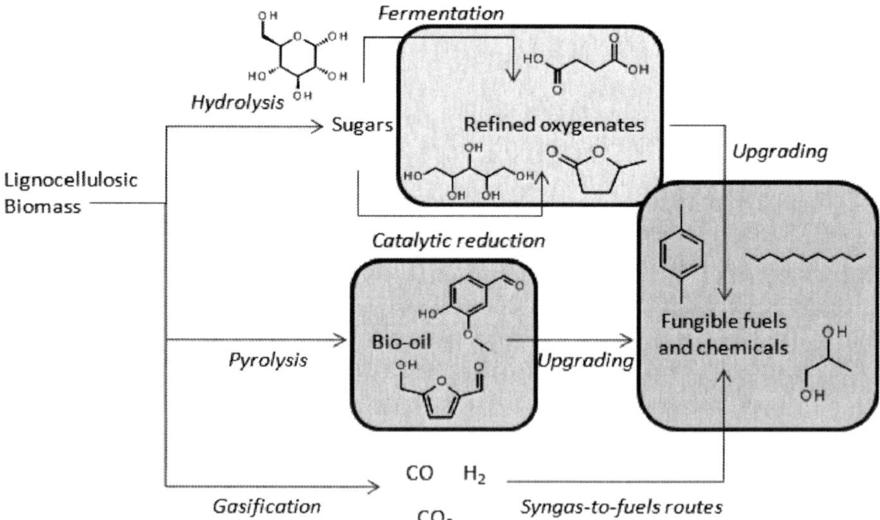

Figure 3.1 Summary of key biorefining routes. The boxes indicate example species of interest for surface science investigations on metal surfaces.

feedstock flexibility, since any $C_xH_yO_z$ feedstock can in principle be gasified. In fact, although the high oxygen content of biomass presents some unique challenges for gasification and subsequent upgrading, the basic outlines of many processes have been explored extensively for fossil fuel (e.g. coal) processing.[29] For this reason, the current review will not focus on catalytic technologies related to gasification, despite their industrial promise for biofuels applications.

3.3.2 Catalytic Pyrolysis and Catalytic Upgrading

Pyrolysis of biomass is a similar process to gasification, except that the temperature of the deconstruction reaction is lower.[30–33] The lower temperature causes the degradation of lignocellulosic biomass to be less complete, i.e. to not proceed all the way to synthesis gas. The principal products of biomass pyrolysis are some light gases and 'bio-oil', a complex, unstable mixture of biomass-derived oxygenates. Cellulose pyrolysis tends to produce a mixture of oxygenated chemicals, including levoglucosan (commonly considered the initial decomposition product) and furanic compounds.[34] Pyrolysis of lignin produces a variety of phenyl and phenolic compounds that can subsequently be catalytically upgraded and employed in gasoline blends.[35] While somewhat similar in appearance to crude oil, this bio-oil is generally not suitable as a widely applicable fuel or oil refinery feed. Instead, various upgrading reactions are needed to convert the bio-oil into useful fuels and chemicals. These upgrading reactions generally require the use of heterogeneous catalysts. Furthermore, recent investigations of the pyrolysis step itself have focused on catalytic pyrolysis, which has the potential to produce a higher-quality bio-oil prior to upgrading steps.[36] As summarized in Figure 3.1, pyrolysis pathways involve a variety of aromatic and oxygenated intermediates; a focus of this review is identifying insights from mechanistic studies of related intermediates on catalyst surfaces.

3.3.3 Hydrolysis of Cellulosic Biomass

Another key deconstruction technology is biomass hydrolysis, in which the ether linkages of cellulose and hemicellulose are hydrolyzed to produce individual sugar molecules. As discussed below, these sugars can serve as feedstocks for a variety of upgrading reactions. A large number of catalysts (including treatment with mineral acids and enzymes) have been identified for saccharification.[37] A smaller number of studies have focused on identifying heterogeneous catalysts capable of catalyzing hydrolysis reactions.[38,39] A primary advantage of the application of heterogeneous catalysts would be the ability to recover the catalyst, since recovery of the expensive enzymes is a major issue. Heterogeneous catalysts could also potentially be employed at elevated temperatures to increase the hydrolysis rate. Problems with mass transfer limitations to the catalyst and coking of the catalyst may be significant, however.

3.3.4 Aqueous Phase Processing of Sugars

Once the monosaccharide sugars are liberated from biomass through the processing described above, the C_5 and C_6 sugars can be converted to fuels and chemical products via a range of different processes conducted in the aqueous phase.[40,41] Here, aqueous phase processing is commonly employed because the highly oxygenated compounds derived from biomass have low volatility. Sugars can be hydrogenated to polyols such as sorbitol and xylitol on supported metal catalysts. By selecting appropriate catalysts and operating conditions, the polyols can further undergo a variety of reactions. For example, reforming of polyols leads to production of hydrogen and CO_2 or to synthesis gas through multiple C–C bond breaking steps.[40,41] Addition of hydrogen (or generation of hydrogen through reforming) can alternatively be used to drive dehydration/hydrogenation reactions that produce hydrocarbons and other deoxygenated compounds such as hydroxymethylfurfural (HMF) and related compounds.[42,43] HMF represents a versatile compound for subsequent refining, with direct hydrogenation leading to chemical products and aldol condensation reactions followed by dehydration/hydrogenation leading to higher molecular weight alkanes. This chemistry is discussed in greater detail below. A large number of processes requiring selective reactions of numerous functional groups (olefin, alcohol, aldehyde, ketone) has been investigated in the literature, with recent reviews providing detailed pathways.[40,41] Some example pathways from sugars to fuels are also described below.

3.3.5 Upgrading of Fermentation Products

Sugars can also be deoxygenated to a variety of platform chemicals and fuels using microbes in fermentation reactions. An enormous effort is devoted to developing organisms and reaction schemes for these fermentations. In many cases, fermentation provides alternative routes to partially deoxygenated compounds that serve as useful feedstocks for heterogeneous catalysis. Compounds such as the 4-carbon diacids (succinic, maleic, and fumaric acid), 3-hydroxypropionic acid, itaconic acid, and many others can be produced through fermentation routes.[44,45] Upgrading of these compounds to other fuels and chemicals represents a potentially attractive approach given that biological systems are highly evolved for reactions of carbohydrates, but complete deoxygenation results in accumulation of compounds that are often highly toxic to the microorganism.[46] Therefore, metal-catalyzed reactions that lead to deoxygenation of chemical intermediates produced by fermentation have been the subject of extensive investigation, as highlighted by examples below.

3.4 Surface Science Methodology

A very large number of techniques have been developed for understanding mechanisms of surface reactions on metals and other materials. Many surface science investigations are conducted at ultrahigh vacuum (UHV) on single crystals or other model materials. Thus, two key approximations are made in using such catalysts to study reaction mechanisms important for catalysts. The first approximation is that the low pressures used in many surface science experiments are adequate for probing catalytic chemistry that occurs at much higher pressures in application. Low pressures are generally needed to maintain clean metal surfaces and to permit the use of electron-based spectroscopies (though in the last few years high-pressure electron spectroscopy has become possible).[47] The second approximation is that a 'simple' surface, such as a metal single crystal, can be used to understand chemistry that occurs on polycrystalline materials with many defects, with polydisperse metal crystallite sizes, and with many potential promoters, poisons, and contaminants. Here, single-crystal surfaces are often employed so that one is using a relatively well-understood model substrate with a controllable surface topography. To the extent that these approximations do not conform to the *in situ* conditions of catalysis, they are referred to as the 'pressure gap' and 'materials gap', respectively.[48] One must always bear in mind that such gaps can result in apparent discrepancies between surface science experiments and the observations of catalysis. Yet, as discussed above, a number of surface science-based investigations have in fact yielded an ability to design technical catalysts. Furthermore, identification of reasons for pressure or materials gaps has led to a much better understanding of the mechanisms of key reactions.[49] While keeping in mind the approximations of surface science, one must also acknowledge the chief benefit: that one can generally gain a much more thorough understanding of surface reaction pathways using such model studies. In other words, the surface science approach does not necessarily yield a *simulation* of surface behavior during catalysis, but can provide a foundation for understanding surface reaction mechanisms.

Several types of surface science investigations are employed for determining surface reaction mechanisms. An overview of the different types of techniques is provided in Figure 3.2. One key method is temperature-programmed desorption (TPD). In this method, the temperature of an adsorbate-covered sample is ramped at a linear rate, and a mass spectrometer or some other gas detection technique is employed to measure desorption product yield as a function of temperature and time. For the case of simple desorption of an adsorbate, TPD provides information on the adsorbate–surface binding strength, which increases with increasing desorption temperature, and on the coverage of adsorbate, which scales with peak area.[50] However, TPD can also be used to detect volatile products that result from reactions of adsorbates; in this implementation, TPD is often called temperature-programmed reaction spectroscopy (TPRS).[51] Because TPD provides highly quantitative information, it is a useful tool for measuring the kinetics of reactions producing volatile products, and is a vital tool in evaluation surface reaction mechanisms.

While TPD provides a key tool for studying surface kinetics, detailed studies rely on the ability to directly probe the adsorbed state of molecules on surfaces. In much of the work discussed below, the adsorbed states of such intermediates are determined by vibrational spectroscopy, in particular with reflection

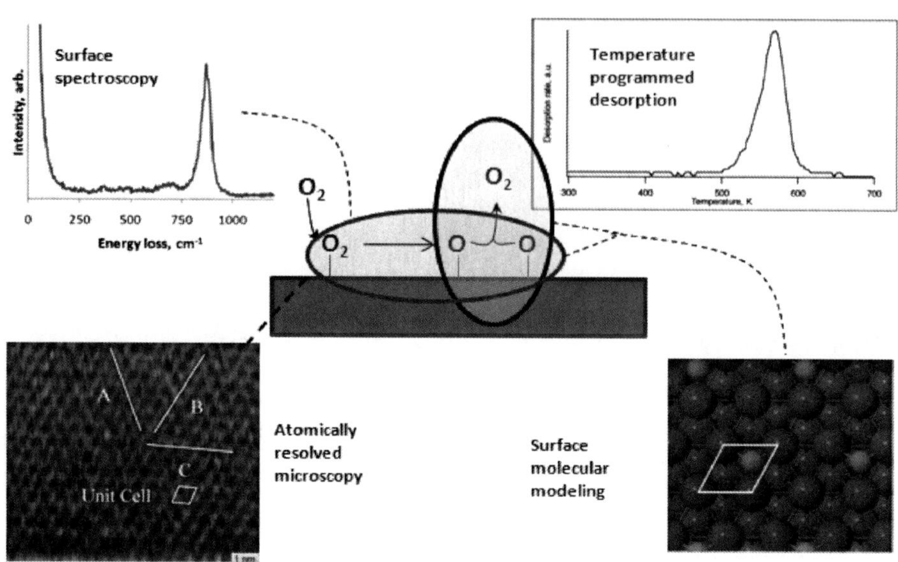

Figure 3.2   Principle techniques of the surface science techniques used in these investigations, using oxygen adsorption, dissociation, and recombination as an example chemistry. Adsorbed O_2 is detected using spectroscopic techniques such as high resolution electron energy loss spectroscopy (HREELS), shown at top left for O_2 adsorbed on Pt(111) at 100 K. The disappearance of the molecular O–O stretch and appearance of the surface–O stretch (not shown) indicates dissociation of O_2 to form adsorbed O atoms. Heating of the surface results in recombinative desorption of O_2 from metal surfaces (top right panel for $O(s) + O(s) \rightarrow O_2(g)$ on Ag(110)) which can be monitored with TPD. Microscopies such as STM can be employed to monitor surface structure; the lower left panel shows a Pt(111) surface covered by 0.3 ML of O atoms.[185] Finally, all these processes can be modeled using molecular modeling approaches (0.25 ML O covered Pt(111) shown at bottom right) such as density functional theory.

absorption infrared spectroscopy (RAIRS) and high resolution electron energy loss spectroscopy (HREELS).[52] Both methods are based on measuring energy losses from a monochromatic beam scattered off a surface; these losses correspond to the vibrational energy (frequency) of different stretching and deformation modes of surface species. Another commonly employed surface spectroscopic method is X-ray photoelectron spectroscopy (XPS), which measures the energy of core electrons of surface atoms as a probe of the local chemical environment.[53] Using such measurements one can monitor, for example, when a C–O bond is broken so that both the C and O atoms become bound to the metal surface. A variety of related electron spectroscopies, such as near edge X-ray absorption fine structure (NEXAFS) and ultraviolet photo-electron spectroscopy (UPS) also provide information on the local environment of surface atoms.[53,54]

The techniques described above provide a great deal of information on the chemical environment of the surface, but microscopy is needed to analyze the detailed arrangement of surface atoms. While imaging that allows the detailed interpretation of reaction pathways for adsorbates is still difficult, techniques such as scanning tunneling microscopy (STM) and atomic force microscopy (AFM) allow one to gain a real-space picture of how surface reactivity depends on the detailed atomic structure of surfaces. For example, STM has been used to investigate the role of step/defect sites in a variety of surface reactions.[55]

Finally, quantum-based simulation methods have become tremendously important in surface science approaches to catalysis.[8] The principle quantum mechanical method used for such studies is density functional theory (DFT) because it allows accurate investigation of trends in surface chemistry at relatively low computational cost. A very useful feature of DFT calculations is that they can be employed both to identify stable surface intermediates and to calculate simulated results to all the types of experiments described above. For example, DFT can be used to simulate transition state energies that can be compared to information from TPRS,[56] vibrational spectra for adsorbed intermediates,[18] shifts in electronic orbital energies for comparison to XPS[15] or NEXAFS,[57] and STM images for a defined surface structure.[58] Direct comparisons between experimental and theoretical spectra can then be employed to refine the mechanistic picture of the surface chemistry. Such an approach has been demonstrated by the cited studies and many others to provide a high level of detail in understanding surface chemistry on metal surfaces.

The work described below has generally used combinations of the techniques described above to identify surface reaction mechanisms. In most cases, we will focus on the surface intermediates and reaction mechanisms themselves rather than the characterization techniques that were used to identify the mechanisms. The reader is referred to the cited references for details on the identification.

3.4.1 Adsorption and Reaction of Key Functional Groups on Metals

The adsorption and reaction of simple molecules with a key functional group (e.g. ethylene, methanol, formaldehyde, formic acid) on metal surfaces has been probed in thousands of investigations over several decades. The purpose of the following sections is therefore not to provide a comprehensive review of each type of adsorbate, but rather to provide broad indications of how these key molecules bind and react on metal surfaces. The reactions of these relatively simple adsorbates will then be used to guide analysis of more complex adsorbates. A cartoon description of the key surface intermediates described below can be found in Figure 3.3.

3.4.2 Olefins

Olefins such as ethylene have been among the most extensively studied molecules in surface science investigations of catalysis.[59–63] Only a brief

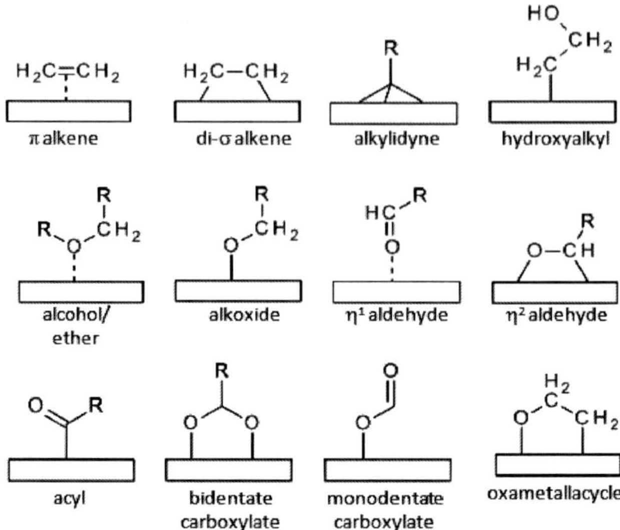

Figure 3.3 Key intermediates observed in reactions of olefins and oxygenates on metal surfaces.

overview is provided here. Ethylene generally binds to metal surfaces in one of two ways: through an interaction of the ethylene π electrons with the surface (dubbed the π adsorption state), or through formation of sigma bonds with each of the carbon atoms (referred to as the di-σ adsorption state). Both states can be formed on surfaces such as Pd, Pt, and Ni, with the more stable di-σ state typically being formed preferentially at low coverage and the more weakly adsorbed π state being formed at higher coverage. Although the π state is bound less strongly, it has been determined to be the more reactive species for ethylene hydrogenation.[64,65] Whereas the π adsorption state typically desorbs intact from metal surfaces, the di-σ state can either desorb or undergo further decomposition reactions. On Group VIII metals such as Pt and Pd, ethylene undergoes dehydrogenation–rehydrogenation to produce highly stable ethylidyne intermediates (which are notable spectator species present during steady state olefin hydrogenation). On metals to the left of the Group VIII metals such as Fe, ethylene can also undergo C–C scission to produce adsorbed CH_x intermediates.[66] These intermediates typically decompose at high temperature to produce H_2 and surface carbon. On Group IB metals, ethylene tends to adsorb very weakly through a π mode and to desorb without reaction.[67,68]

3.4.3 Alcohols

The reactivity of alcohols and other oxygenates has been studied extensively on a variety of transition metals, and only a brief overview is provided here. The reader is referred to more extensive reviews for additional details.[69,70] On most catalytic metal surfaces, alcohols adsorb through their oxygen atom and decompose via a low activation barrier O–H scission to form stable alkoxide ($-O-CR_3$) intermediates.[71–88] Formation of alkoxide intermediates must be assisted by predosed surface oxygen for Ag surfaces, and by either surface oxygen or elevated dosing temperature on Cu.[89–91] On many surfaces (Pt, Pd, Ni, Ru), the alkoxide intermediates further undergo α-H elimination to form surface aldehyde intermediates.[82,84,87,92–97] Such intermediates may be bound to the surface either through both their C and O atoms in an η^2 arrangement, or through the oxygen atom only via an η^1 orientation (see Figure 3.3). The preference of one form over the other is dictated by both the surface conditions (e.g. coadsorption of oxygen) and the molecular structure of the alcohol. Whereas the more weakly bound η^1 intermediates tend to desorb without reaction, the more strongly adsorbed η^2 aldehydes decompose through an acyl intermediate to produce CO and hydrogen, which desorb from the surface at higher temperatures.

Decomposition of alkoxides has been proposed to proceed somewhat differently on other surfaces. For example, on Rh(111) surfaces ethoxides have been proposed to decompose via β-H elimination to produce transient surface oxametallacycle intermediates, with aldehyde intermediates being conspicuously absent.[98] The oxametallacycle undergoes decomposition to CO, hydrogen, and surface hydrocarbon fragments. On coinage metal surfaces such as Cu and Ag, alkoxides can undergo three reaction pathways: α-H abstraction to aldehydes, reaction with adsorbed aldehydes to produce carboxylate species through ester intermediates, and hydrogenation to produce alcohols.[89–91] Decarbonylation pathways are not active, as C–C bond cleavage is very difficult on the coinage metals. Overall, the tendency of alcohols to form alkoxides, and for primary and secondary alkoxides to decompose through C–H scission (often through carbonyl intermediates, i.e. alcohols and ketones) has been observed on numerous surfaces, and represents an important pathway for understanding reactions of alcohols.

It is also worth noting that some theoretical studies have suggested that hydroxyalkyl species are more stable adsorbed isomers than alkoxide species, though the barriers for formation of these species are thought to be higher.[92] Interestingly, the barriers for bond scission in hydroxyalkyl intermediates, once formed, have been proposed to be lower compared to alkoxides.[99]

3.4.4 Aldehydes and Ketones

The surface chemistry of aldehydes and ketones has also received considerable scrutiny, and as discussed above aldehydes are key intermediates in many alcohol decomposition mechanisms. On many Group VIII metals important for catalysis, adsorption of carbonyl compounds in both η^1 and η^2 geometries can be observed.[71,100–104] On Cu and Ag, however, the η^1 geometry is formed exclusively.[105] This has been explained in terms of the reduced availability of surface d electrons for backbonding into the C=O π^* system in Group IB metals; furthermore, this reduced availability has been used to explain the preference for formation of η^1 adsorbed aldehydes on Group VIII surfaces precovered by electron-withdrawing adsorbates such as oxygen.[106]

Interestingly, recent DFT calculations for acetaldehyde adsorption on the Pt(111) surface suggest that adsorption through the aldehydic hydrogen represents the most stable structure at high coverage.[107] As discussed above, η^2 carbonyl species typically react on transition metal surfaces via C–H scission at the carbonyl position and subsequent CO abstraction; on some surfaces, the decarbonylation reaction appears to be preceded by dehydrogenation at the β-C position which competes with C–O scission.[104] On Group 1B metals, adsorbed carbonyl compounds generally desorb intact or may react with other adsorbates, but do not undergo decarbonylation.[105] Longer-chain aldehydes generally undergo comparable decomposition reactions; for example, propionaldehyde has been observed to coordinate through an η^2 state on Rh(111) and decompose to CO and ethyl intermediates.[102]

3.4.5 Ethers and Epoxides

Ethers tend to adsorb weakly on transition metal surfaces through donation of a lone pair from the oxygen atom to the surface. The most extensively studied ethers are epoxides, both because of their importance in industrial reactions and because of their higher reactivity associated with strain on the epoxide ring. On surfaces such as Pt, Ag, and Cu, epoxides adsorbed at low temperature (<200 K) desorb without reaction.[108–111] On metals such as Pd and Rh, however, epoxides undergo ring-opening reactions to rapidly produce surface CO and hydrocarbon fragments.[112–114] On Ag surfaces epoxide ring-opening reactions can be induced by dosing the epoxides to the surface at higher temperatures (>200 K) to form stable surface oxametallacycle intermediates.[56,108] The oxametallacycles react via the reverse reaction to produce epoxides upon heating, but also produce other products such as aldehydes.

Studies of other ethers are far more limited. It has been observed that ethers can decompose on catalyst surfaces at high temperatures to produce alkoxides through C–O bond cleavage.[115] On the Rh(111) surface, for example, dimethyl ether adsorbs weakly, but coadsorption of K strengthens adsorption and allows the observation of decomposition to CO and hydrogen through initial C–O scission.[116]

3.4.6 Carboxylic Acids and Esters

The surface chemistry of carboxylic acids, particularly formic acid, has been studied in considerable detail on many metals. On most metal surfaces, carboxylic acids adsorb molecularly at low temperature, with strong H-bonding interactions between neighboring adsorbates.[117,118] Carboxylic acid species generally undergo O–H scission at low temperatures to produce strongly adsorbed carboxylate species. The reactions of formate, produced from formic acid, are surface dependent. On surfaces such as Cu, Ag, and Pt, formation of CO_2 and hydrogen is preferred.[119–123] On other surfaces, such as Pd, Ni, Ru, and Rh,[117,118,124–126] CO formation is additionally observed. On certain metal surfaces, such as Ru, a monodentate formate state is also observed, which is converted to bidentate at higher temperatures.[118] Bidentate formate can decompose via C–H scission to produce CO_2 or by C–O scission to produce CO. Longer chain carboxylic acids, such as acetic acid, generally adsorb and decompose via similar mechanisms (i.e. through acetate intermediates) on metal surfaces.[127–129]

Studies of the reactions of simple esters are far rarer. The thermal chemistry of the cyclic ester γ-butyrolactone have been investigated on the Pd(111) and Pt(111) surfaces.[130] On both surfaces, weak adsorption through the ring O atom has been suggested. The adsorbed ester primarily desorbs without significant reaction on Pt, but scission at both C–O bonds (to produce CO and CO_2, depending on which bond is broken) is detected on Pd(111). Decarbonylation is observed to be the primary pathway. A recent DFT study of the reaction of methyl acetate on Pd(111) similarly shows a weakly adsorbed molecular state that can undergo dehydrogenation at the methyl function followed by scission of C–O bonds, with the decarbonylation pathway preferred.[131]

3.4.7 Summary of Adsorption and Reaction Trends

The studies of the monofunctional adsorbates discussed above provide a template for understanding reaction of multifunctional adsorbates. While such reactions are considered explicitly below, it is worth pointing to some general trends. Adsorbate–surface interactions that involve the breaking of a bond within the adsorbate tend to be strong, whereas those that do not tend to be weak. Thus, for example, adsorption energies of intermediates such as alkoxy, hydroxyalkyl, η^2 aldehyde, di-σ alkene, and formate are strong, whereas adsorption strengths of acids, alcohols, esters, ethers, η^1 aldehydes, and π alkenes are weak. Thus, one may expect that trends in binding of multifunctional oxygenates will depend on the identity of the individual functional groups, such that (for example) an unsaturated ether would preferentially bind in an intact state through di-σ adsorption of the olefin function. As discussed below, this in fact turns out to be the case. One also notes trends across the periodic table, with the Group IB metals exhibiting qualitatively different behavior than the Group VIII metals. The low energy of d electrons in the IB metals makes strong adsorption through bond-breaking reactions far more difficult, especially for C–C and C–O dissociation reactions. Thus, metals from Group IB can serve as key modifiers for the reactivity of Group VIII surfaces, as will be explained in greater detail below.

3.5 Reactions of Multifunctional Oxygenates on Metals

Note that the reactions of single functional groups can sometimes provide a picture of what should happen.

Many biomass-derived intermediates contain multiple functional groups, and numerous biofuels and biochemicals production routes depend on the selective reaction of a single functional position. While studies of the reactions of individual functional groups, as discussed above, can provide some insight, it is important to also conduct model studies incorporating multiple functional groups. These studies are considerably less extensive than the investigations of simple probe molecules described in the previous section. However, they do illustrate several key points regarding reactions of multifunctional molecules, as discussed below.

3.5.1 Unsaturated Oxygenates

Perhaps the most-studied multifunctional oxygenates are molecules that contain an oxygenate functionality and an unsaturated C=C bond. The interaction of α,β-unsaturated aldehydes (in particular, acrolein, crotonaldehyde, and prenal) with metal surfaces has drawn particular attention in the literature because of their importance in producing high value enols. The most common reaction pathways observed in such hydrogenations are summarized in Scheme 3.1. Many researchers have investigated catalysts for increasing the selectivity for hydrogenation of the aldehyde function over olefin hydrogenation or decarbonylation, in particular the modification of Pt-group catalysts with 'oxophilic' metals such as Sn and Fe. Experimental studies on single crystal surfaces of the common hydrogenation metals Pt,[132–134] Pd,[135–137] and Rh[138] have generally suggested that α,β-unsaturated aldehydes adopt a flat-lying adsorption geometry at low coverages, such that both the olefin and aldehyde functional groups coordinate with the surface. However, the detailed surface chemistry is highly dependent on the surface composition, adsorbate coverage, and the molecular structure of the α,β-unsaturated aldehyde. For example, acrolein has been found experimentally to preferentially bind through both functional groups in an η^4 configuration on Pd(111) and Pt(111), and produces a variety of decomposition products during temperature-programmed desorption.[133,136] In contrast, irreversibly adsorbed crotonaldehyde has been found to interact preferentially through the olefin function, as evidenced by a high-intensity C=O stretching feature in vibrational spectroscopy experiments.[133,137] The main reaction pathway in all cases, however, appears to be decarbonylation. In general, methyl substituent effects at the olefin position exhibit a significant effect on the adsorbate phases on the surface.[139]

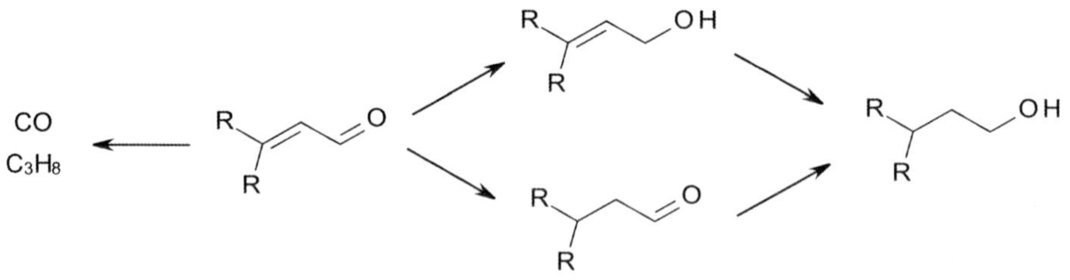

Scheme 3.1 Pathways for hydrogenation of α,β-unsaturated aldehydes on metal catalysts. All steps above require addition of one molecule of H_2.

More recent studies employing both surface science experiments and density functional calculations indicate that the adsorption chemistry of unsaturated aldehydes is more complex than originally envisioned, with multiple intermediates of different degrees of coordination – from relatively upright adsorbates bound through one of the two functional groups, to flat-lying adsorbates – existing on the surface under various coordination conditions.[140–145] Model studies furthermore indicate that addition of oxophilic metals such as Sn and Fe to the surface causes preferential upright adsorption through the aldehyde function, leading to higher selectivity to unsaturated alcohols.[140,141,146,147] Recent studies have suggested that while Pt catalysts typically show higher selectivity for forming unsaturated aldehydes, H addition to the oxygenate function has a lower inherent barrier, with the transition state for alcohol formation stabilized by a precursor state (Figure 3.4).[142] Formation of the unsaturated aldehyde is proposed to be more selective because of a lower barrier to desorption of the aliphatic product.

While α,β-unsaturated aldehydes represent the most extensively investigated adsorbate system, unsaturated alcohols, epoxides, esters, and ethers have also been the subject of some investigations. The adsorption of 1-epoxy-3-butene (EpB) has been investigated on the Ag(110),[18,56,57] Pt(111),[148] and Pd(111)[137] surfaces (see Scheme 3.2). Not surprisingly given the comparably weak adsorption of epoxides, the most stable mode of adsorption is through the olefin function in a di-σ configuration on the Pd and Pt surfaces. On Ag, which has weak binding proclivity for the olefin function, EpB is weakly adsorbed through the epoxide at low temperatures; however, for higher adsorption temperatures (~300 K), EpB undergoes ring-opening to reversibly form a highly stable oxametallacycle intermediate on Ag(110) and Ag(111). On the (111) surfaces of Pd and Pt, however, adsorbed EpB undergoes ring-opening at approximately 200 K to form an adsorbed unsaturated aldehyde, which undergoes decarbonylation to CO, propylene, and surface carbonaceous species by approximately 300 K. Thus, while adsorption occurs through the olefin function, decomposition occurs through the unsaturated oxygenate function. The detailed understanding of EpB surface chemistry enabled the design of a bifunctional supported AgPt catalyst for improved selectivity to olefin hydrogenation.[149] In this catalyst, the Pt serves as the hydrogenation site for reaction of the olefin function, whereas the Ag appears to selectively bind to

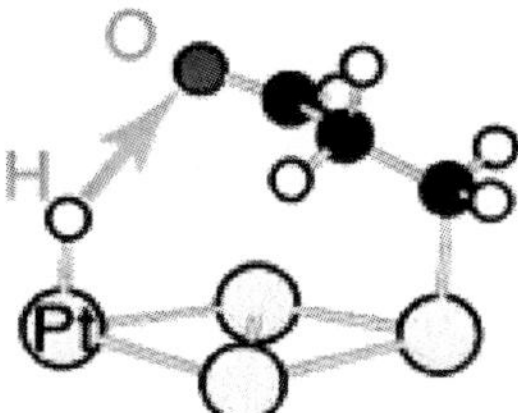

Figure 3.4   Reaction path for selective aldehyde hydrogenation in crotonaldehyde, taken from ref. 142.

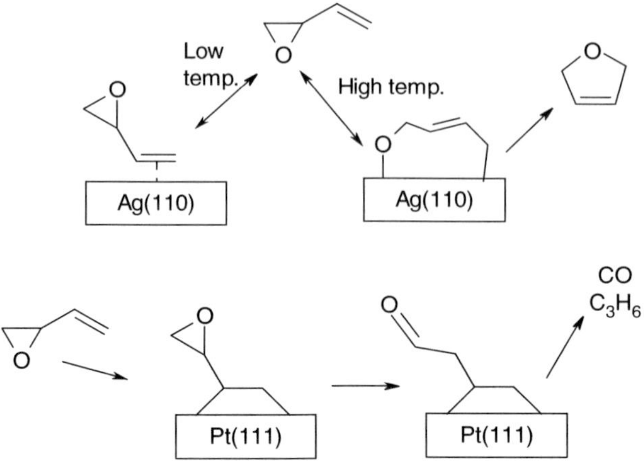

Scheme 3.2 Adsorption and reaction pathways for 1-epoxy-3-butene on Ag(110) and Pt(111).

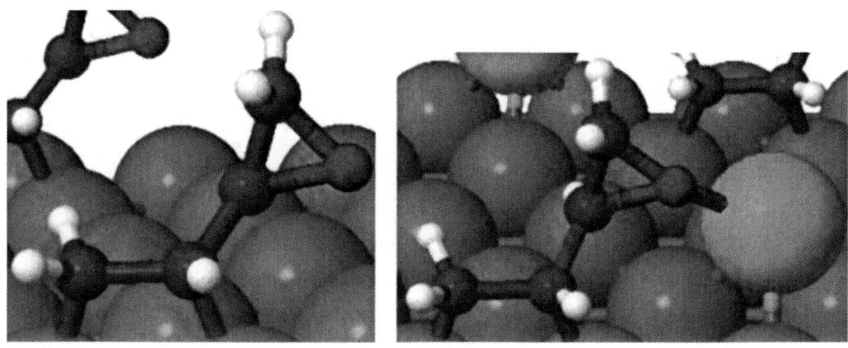

Figure 3.5 DFT-optimized geometry of 1-epoxy-3-butene on Pt(111) and on a Pt(111) surface covered by 1/9 monolayer of Ag atoms.[149]

and 'protect' the epoxide function. The preferential coordination of the epoxide function (Figure 3.5) appears to help prevent decomposition of the epoxide ring to an aldehyde.

The adsorption of other cyclic unsaturated ethers, in particular furans, has also been studied on a few surfaces, but especially on Pd(111). Please see Scheme 3.3 and Figure 3.6 for a summary of this chemistry. Furan, with two unsaturated C=C bonds, has been found to strongly adsorb through all four carbon atoms.[150–154] At higher temperatures (>300 K), furan reacts via ring-opening and decarbonylation to produce CO and a surface C_3H_3 species. This C_3H_3 intermediate subsequently undergoes coupling to produce surface-adsorbed benzene.[154] The singly unsaturated dihydrofuran (DHF) reagents 2,5-DHF and 2,3-DHF are also found to adsorb through their olefin function, and to react via dehydrogenation to produce adsorbed furan above 200 K, which undergoes the decomposition channels described above.[155] Additionally,

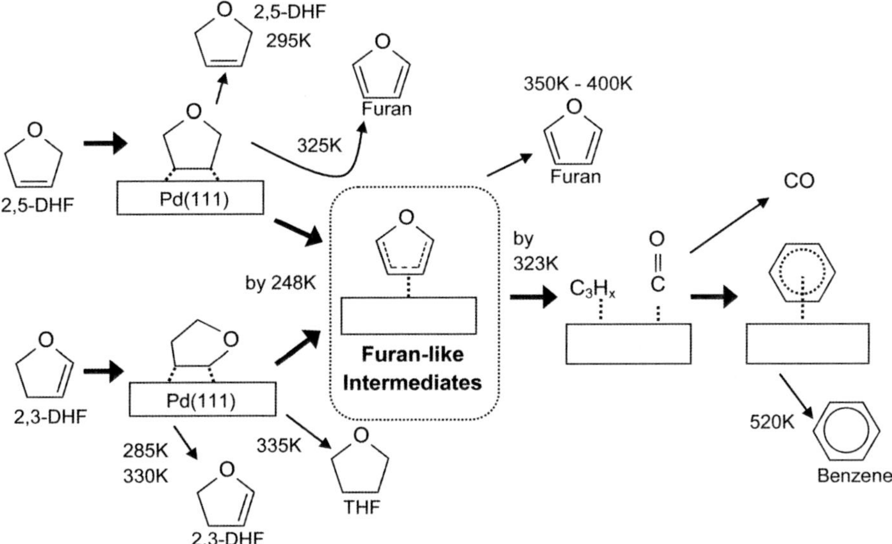

Scheme 3.3　Partial reaction pathways for 2,5-dihydrofuran, 2,3-dihydrofuran, and furan. For simplicity, evolution of hydrogen is not shown. Note that these cartoon images do not accurately represent the geometry of a di-σ adsorbed olefin species; for a more accurate depiction, please see Figure 3.6.

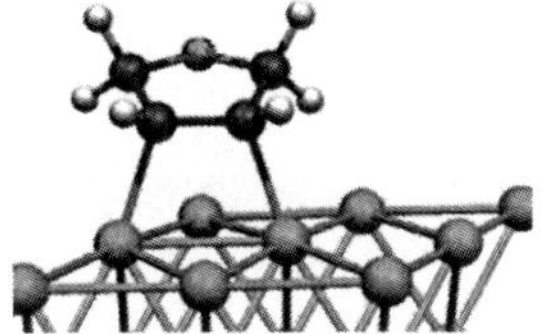

Figure 3.6　Optimized adsorption geometry of 2,5-dihydrofuran.[155]

2,3-DHF undergoes hydrogenation to tetrahydrofuran (THF), which is not observed for 2,5-DHF; however, for 2,5-DHF there is an additional low-temperature pathway that produces furan. The results suggest that the proximity of the C–O–C function plays a key role in the reactivity of the adsorbed molecule. However, the generally higher activation barrier for ring-opening of furans compared to epoxides indicates the critical role of ring strain in ring-opening reactions of ethers.

The adsorption of 2-propen-1-ol (allyl alcohol) has also been studied on a few surfaces, including Ag, Cu and Rh. On Ag(110), allyl alcohol adsorbs on the O-covered surface to form an unsaturated alkoxide (Scheme 3.4(a)), which is oxidized to acrolein following mechanisms similar to those described above for simple alkoxides.[156] On the Cu(110) surface, 2-propenol initially reacts via dissociation of the O–H bond to form an alkoxide species that may exhibit a weak π interaction of the olefin function (Scheme 3.4b), owing to the slightly

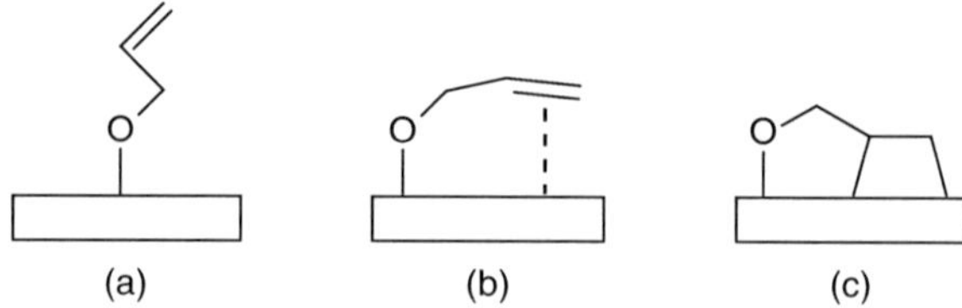

(a)   (b)   (c)

Scheme 3.4 Alkoxy intermediates observed in reactions of allyl alcohol on metal surfaces: (a) simple alkoxy, (b) alkoxy with π olefin, (c) η^3 adsorbed alkoxy.

higher affinity of the olefin for Cu surfaces.[157] As in other adsorbed alkoxides, this species can undergo dehydrogenation to form the α,β-unsaturated aldehyde acrolein, but due to the olefin–surface interaction can also undergo hydrogen rearrangement to produce surface oxametallacycle intermediates that can subsequently produce saturated alcohols and aldehydes. The oxametallacycle production pathway represents a new pathway for alkoxides introduced by the olefin function, much like the additional pathway opened up to epoxides on Ag(110) for EpB. In contrast, on Rh(111) the stronger affinity for the olefin function results in low-temperature adsorption in an η^3 configuration (Scheme 3.4c), i.e. forming simultaneous alkoxide and di-σ olefin linkages. This highly coordinated species undergoes decarbonylation at low temperatures, but does not proceed through an aldehyde intermediate.[138]

The reactions of the unsaturated esters vinyl acetate (on Pd(111))[158] and 2(5H)-furanone (on the (111) surfaces of Pd and Pt)[159] have also been investigated. Consistent with the weak adsorption of the ester functionality, these molecules are again found to adsorb through their olefin function in the monolayer state. As the surface is heated, vinyl acetate undergoes scission at both the C(=O)–OC and C(=O)O–C positions to produce a range of products. In contrast, 2(5H)-furanone undergoes ring-opening almost exclusively at the C(=O)–OC position, suggesting that the constraints of the cyclic ester structure alter selectivity compared to the acyclic vinyl acetate adsorbate.[159] On Pt(111), scission at both positions is again preferred, but with a higher selectivity toward C(=O)–OC scission. Whereas rapid double decarbonylation is observed upon ring opening of 2(5H)-furanone on Pd(111), the resulting multiply coordinated ring-opened intermediates – aldehydic oxametallacycles and unsaturated carboxylates, respectively – are stable on Pt(111) up to at least 400 K. Also, as shown in Scheme 3.5, the reactivity of unsaturated esters varies markedly from their saturated counterparts: compared to γ-butyrolactone, 2(5H)-furanone adsorbs more strongly on Pd(111), decomposes at a higher temperature, cleaves at the C(=O)–O position more selectively, and forms different types of carbonaceous decomposition products.

In summary, unsaturated oxygenates tend to exhibit binding through their olefin function in a di-σ configuration on Group VIII metal surfaces, particularly at low coverages. Multiply coordinated structures such as η^4 adsorbed unsaturated aldehydes may also be formed if bond breaking reactions such as rehybridization can occur, but intact adsorption of alcohols, ethers, and ester function on the surface appears to be disfavored relative to di-σ olefin adsorption.

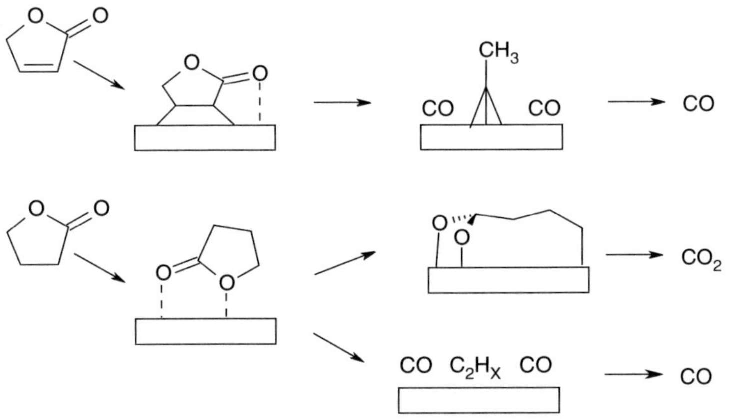

Scheme 3.5 Summary of reaction pathways of 2(5*H*)-furanone and γ-butyrolactone on Pd(111).

3.5.2 Polyols

Polyols (or 'alcohol sugars') such as glycerol, xylitol, and sorbitol are commonly studied feedstocks for aqueous phase processing. Diols including ethylene glycol (EG), 1,2-propanediol (PDO), 1,3-propanediol, 1,4-butanediol, and 2,5-pentanediol are valuable chemical products that have been targeted for production form biorenewable feedstocks.[44] Most of the detailed surface-level investigations of polyol chemistry have been conducted over Ag and Pt (or Pt-based bimetallic) surfaces, and all of these studies have employed ethylene glycol or 1,2-propanediol. Madix and co-workers performed much of the initial work in studying the surface chemistry of EG and PDO on the Ag(110) surface. Consistent with the relatively weak activity of Ag surfaces, EG and PDO desorb without reaction from clean Ag(110).[160,161] As with the simple alcohols described above, the Ag(110) surface must be precovered with O to initiate decomposition reactions of polyols through scission of O–H bonds. On O-covered Ag(110), ethylene glycol decomposes to form ethylenedioxy ($-OCH_2CH_2O-$) at low temperatures ($<180\,K$)[160] (see Scheme 3.6). Ethylenedioxy is stable up to $350\,K$, where it decomposes through a pathway that produces EG along with glyoxal (CHOCHO). The production of glyoxal is preceded by formation of a surface aldehydic alkoxide species ($CHOCH_2O-$).[160]

Surface studies of PDO on Ag(110) reveal similar trends, with PDO decomposing by $215\,K$ on O-covered Ag(110) to produce 1,2-propanedioxy, $-OCH(CH_3)CH_2O-$.[161] Propanedioxy decomposes above $300\,K$ to form the gas phase products acetol (CH_3COCH_2OH), lactaldehyde ($CH_3CH(OH)CHO$), and pyruvaldehyde ($CH_3CHOCHO$), indicating comparable C–H scission kinetics at both the primary and secondary carbon atoms.[161] In the presence of excess oxygen, C–C bond cleavage reactions of propanedioxy are also observed below $300\,K$, producing surface acetate species.[162]

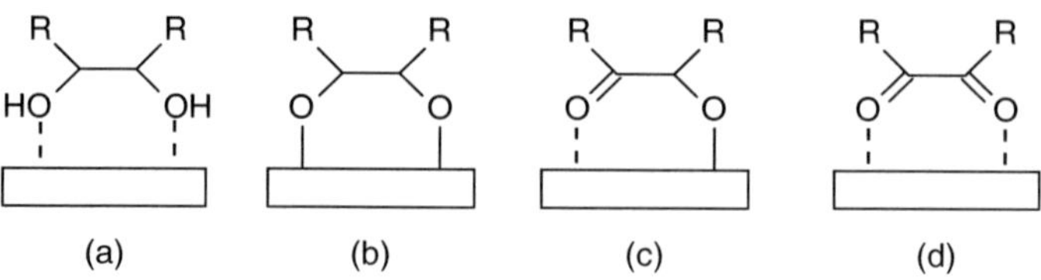

Scheme 3.6 Adsorption and decomposition intermediates of polyols: (a) diol, (b) dialkoxide, (c) aldehydic alkoxide, (d) dialdehyde.

Chen and co-workers have investigated the chemistry of EG on Pt(111), Ni(111), and on different types of PtNi bimetallic surfaces.[163,164] On these more reactive surfaces, EG undergoes complete decomposition to CO and H_2, with some formation of surface carbon.[163] However, the surfaces show clear differences in the extent of EG decomposition. On Pt(111) and on a Pt-Ni-Pt(111) surface in which Ni atoms occupy the second layer beneath at Pt monolayer, the extent of decomposition is quite low, whereas substantial decomposition is detected on Ni-Pt(111) and Ni(111) surfaces.[164] The authors correlated the activity of the surfaces with the position of the center of the d-band relative to the Fermi level, with higher d-band energies corresponding to more reactive surfaces as expected. The decomposition reactions again produce ethylenedioxy, which subsequently dehydrogenates all the way to CO through sequential C–H dissociation steps.[164] Similar trends in decomposition reactivity are observed when glycerol is used as the reagent,[165] or when technical PtNi catalysts are used rather than single crystal model surfaces.[166] Chen and co-workers have also investigated decomposition activities of EG on PtFe and PtTi bimetallic surfaces, and found a similar correlation with the d-band center, such that surface Fe and Ti atoms are necessary to observe substantial EG decomposition.[167]

The decomposition of EG has also been studied on several other surfaces, including Rh(100), Cu(110), and Mo(110). On Rh(100), EG decomposes through ethylenedioxy by around 150 K, although at high coverage some monodentate ($-OCH_2CH_2OH$) species are observed. Ethylenedioxy decomposes through formation of an aldehydic alkoxide intermediate ($CHOCH_2O-$) to ultimately form CO and H_2.[168] On Cu(110), EG, decomposes through ethylenedioxy to form the dialdehyde species glyoxal, similar to the chemistry on Ag.[169] Only Mo(110) favors carbon–oxygen bond scission, forming ethylene with 85% selectivity.[170] For example, on Cu(110) the ethylenedioxy species breaks down to form glyoxal ($(CHO)_2$).

3.6 Relating Surface Studies to Biorefining Catalysis: Case Studies

3.6.1 Reforming of Polyols and Sugars

The reactions of polyols have been studied extensively on a variety of catalysts. Several routes from polyols to fuels and chemicals have been investigated, and

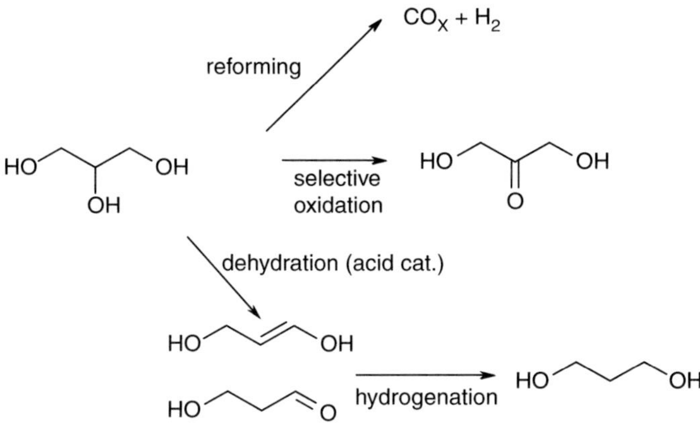

Scheme 3.7 Some key reaction pathways for polyols relevant to catalysis by supported metals.

include (i) reforming to hydrogen and CO or CO$_2$,[40,171] (ii) oxidation to hydroxylated ketones,[172–176] and (iii) dehydration/hydrogenation to oxygenates and alkanes[43] (please see Scheme 3.7). Ring-opening reactions are also of interest for complete reduction and can occur on metal surfaces, though in current practice these occur on acidic or basic materials. Being able to design metal catalysts for high selectivity in the various reaction paths is important for efficient biorefinery operations.

As discussed above, the reaction of diols on Group VIII metals leads to formation of alkanedioxy intermediates; low-temperature α-H elimination leads to formation of dialdehydes, which subsequently undergo C–C scission to produce CO and hydrogen. This pathway is likely responsible for the aqueous phase reforming of polyols described above. On Ag surfaces, activation of the O–H bonds of diols is more difficult, requiring assistance of surface O atoms. Alkanedioxy intermediates can again undergo oxidation to aldehydes under high temperature, but C–C scission does not occur. Thus, one might expect that Group IB metals can be used to modify the surface reactivity of Group VIII metals, since metals such as Ag (and presumably, to an even greater extent, Au) are far less active for O–H, C–H, and (especially) C–C scission reactions. Thus, it can be hypothesized that the role of Au in PdAu and PtAu glycerol oxidation catalysts is to break up ensembles to Pd and Pt atoms to disfavor multiple O–H scission; Pd atoms surrounded primarily by Au might be expected to preferentially undergo a single O–H scission to produce the thermodynamically preferred secondary alkoxide over the primary alkoxide.

The deoxygenation of polyols to alkanes has also been a focus of considerable research.[43] Although catalysts such as supported Ni show reasonable activity for C–O hydrogenolysis, the most promising scheme appears to be the use of a dehydration catalyst (e.g. a solid acid like silica-alumina) together with a hydrogenation catalyst (e.g. Pt). In this case, the metal component serves to hydrogenate unsaturated C=C and C=O bonds that are produced via dehydration. Coordination of such intermediates to surfaces can again occur through several modes: through the hydroxyl, the olefin, and/or the carbonyl. As discussed above, unsaturated oxygenates are generally found to adsorb on Pt and Pd preferentially through the olefin function. In the case of intermediates that produce both an olefin function and a carbonyl function in an α,β-unsaturated carbonyl compound, a variety of adsorption configurations (including an η^4 adsorption structure) can be formed, but most of these again involve adsorption through the olefin function. Tuning of selectivity trends (e.g. controlling the reaction to produce carbonyl compounds for subsequent aldol condensation reactions to increase molecular weight) may require selection of bimetallic catalysts that favor olefin adsorption over oxygenate adsorption, following similar strategies to that used for controlling selectivity in hydrogenation of unsaturated oxygenates described above. For example, the strategy described above that calls for adding Ag to Pt group metals to 'protect' the oxygenate function may be a viable concept in dehydration/hydrogenation schemes aimed at producing oxygenates.

3.6.2 Hydrogenation of Dicarboxylic Acids

The C_4 diacids succinic acid (SA), maleic acid (MA), and fumaric acid (FA) can be produced via fermentation routes, and represent potentially key building blocks in the production of high-volume chemicals.[177,178] For example, hydrogenation of succinic acid can lead to reductive ring-closure of to form γ-butyrolactone, which subsequently can undergo hydrogenation to THF or reductive ring-opening to 1,4-butanediol, a valuable polymerization feed (see Scheme 3.8). All of these steps can be conducted on metal catalysts (e.g. Pd or Ru), and result in the liberation of water. The interactions of diacids with metal surfaces have not previously been studied with surface science techniques. This is likely because of the low volatility of these compounds, which makes them difficult to work with under high vacuum. Based on the well understood mechanism for carboxylic acid chemistry on metals, however, it is expected that diacids would readily dissociatively adsorb on metal surfaces to form dicarboxylate intermediates. The ring-closure mechanism from such intermediates is not clear; however, they may closely resemble the reverse reaction for lactone ring opening to produce carboxylate intermediates that has been observed on Pd(111) and Pt(111) as described above. Once the lactone is formed, hydrogenation to 1,4-butanediol would require selective ring opening at the bond

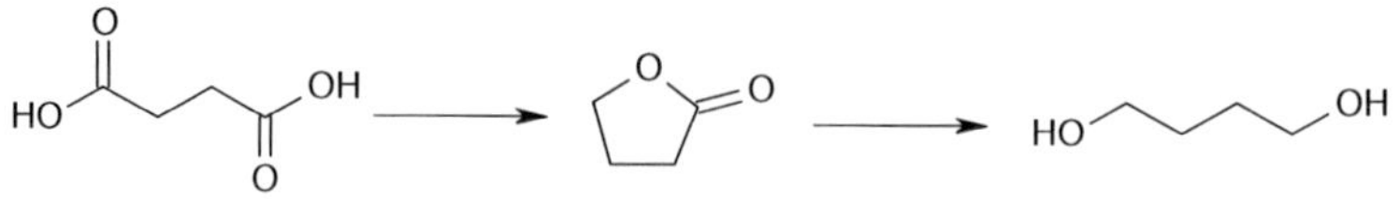

Scheme 3.8 Hydrogenation scheme for production of 1,4-butanediol from succinic acid.

between the carbonyl C atom and the ring O atom; this reaction is observed to occur with high selectivity on Pd(111).

Reactions of unsaturated acids such as fumaric acid may offer interesting challenges and opportunities for production of fine chemicals and connections to surface science. In most investigations reported to date, succinic acid undergoes initial olefin hydrogenation prior to cyclization, but a desirable class of molecules (furanones) could be produced via direct cyclization of fumaric acid with C=C bond retention. Furanones could then potentially undergo ring opening with high selectivity to unsaturated diols on metal surfaces. As discussed above, unsaturated oxygenates tend to bind to Group VIII surfaces through their olefin function; however, this situation can potentially be perturbed by certain bimetallic catalysts that promote oxygenate adsorption and reaction, as those described for hydrogenation of α,β-unsaturated aldehydes above.

3.6.3 Reactions of Hydroxymethylfurfural (HMF)

Hydroxymethylfurfural (HMF) and other furanic species are intermediates in a number of pathways for conversion of biomass to fuels and chemicals, including both aqueous phase processing of sugars and pyrolysis of cellulose.[179–182] HMF is a versatile intermediate, such that valuable products can be produced via a variety of reactions, including selective hydrogenation of either the furan ring or the aldehyde function, selective oxidation of the alcohol function, and aldol condensation for coupling followed by dehydration/hydrogenation to long chain hydrocarbons (see Scheme 3.9). Based on the studies described above, one would expect HMF to coordinate to metal surfaces such as Pd and Pt in at least three stable configurations: (i) through adsorption of the furan ring in an η^4 configuration, which would potentially be predisposed for hydrogenation of the ring or to ring-opening reactions; (ii) through adsorption of the aldehyde intermediate in an η^2 or η^1 configuration such that HMF is predisposed for aldehyde

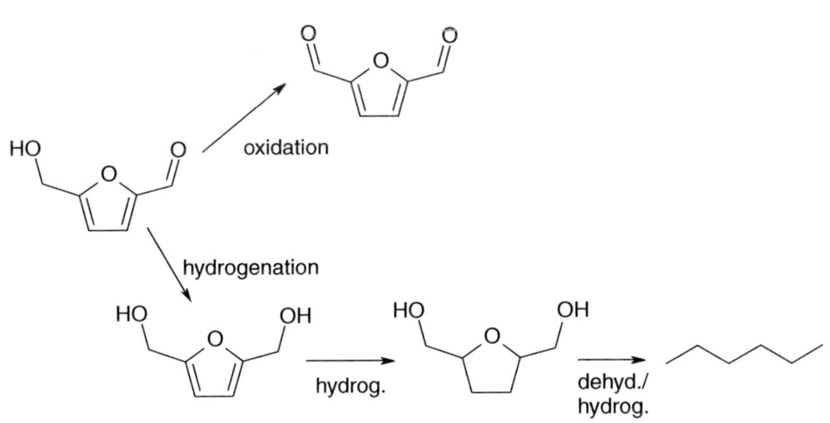

Scheme 3.9 Key reaction pathways of hydroxymethylfurfural (HMF) on metal surfaces.

hydrogenation to an alcohol; and (iii) through dissociative adsorption of the hydroxyl function to produce an alkoxy-type linkage, which is hypothetically the intermediate for oxidation of the hydroxyl to an aldehyde. Furthermore, strong simultaneous adsorption of both the aldehyde and furan functions is likely to be favorable. These high-coordination adsorption modes, while highly thermodynamically favorable, may require large numbers of adjacent sites that are not available under reaction conditions; also, molecules in such configurations may be so strongly adsorbed that they are not sufficiently reactive intermediates.

Catalysts capable of highly selective hydrogenations or oxidations of HMF still need to be developed.[183] The trends discussed above may provide insights into approaches for controlling reactivity. The double bond conjugation of HMF would be expected to lead to intermediates occupying a large number of metal atoms on the surface, such that designing bifunctional surfaces could allow more selective reactions at higher rates. Again, ensemble effects may be important in catalyst design. For example, formation of an alkoxide linkage from the alcohol function would be expected to require only a small ensemble of sites, whereas hydrogenation or ring-opening of the furan ring would be expected to require several contiguous metal atoms due to the high degree of coordination required. Approaches that control the ensemble size of reactive metal atoms may therefore be useful in tuning selectivity for reactions of HMF.

3.7 Summary and Directions of Future Research

The high oxygen functionality of biomass-derived intermediates creates interesting opportunities for conducting chemistry on catalytic metals. Surface science investigations conducted to date have shown how many of the individual functional groups, and some of the functional groups in combination, undergo reactions on key metal surfaces. In most cases the reactions of multifunctional molecules can be understood on the basis of the same kinds of surface–adsorbate interactions that occur for monofunctional adsorbates, but with many details (decomposition kinetics, prevalence of particular adsorption configurations, etc.) dictated by the precise combination of functional groups. While there are excellent opportunities for continued surface science investigations of biorefining intermediates such as more complex polyols, furans, and some phenolics, the low vapor pressures associated with other important intermediates (particularly those involving acid functions) make surface science investigations difficult. Therefore, methods for characterization of surface intermediates and reaction products on model surfaces in the aqueous phase[184] (as well as computational studies) may be of growing importance for understanding the mechanisms of biorefining reactions.

However, the available surface science studies are becoming sufficient to enable catalyst design. Catalyst designs for multifunctional adsorbates may be more complex than for the monofunctional reagents that are more commonly investigated. For example, the bimetallic catalysts (some of which are described above) that have been designed for various reactions of monofunctional reagents have often been designed to have an ideal electronic structure. When multiple functional groups are present in the reagent, it is more likely that a multifunctional catalyst design, where different components of the surface interact with different parts of the adsorbed molecule, will become increasingly important. Of course, controlling the electronic structure of the catalyst is still just as important as for monofunctional reagents, increasing the likelihood that optimized catalysts will be quite complex. Despite the complexity of designing catalysts for reaction of multifunctional reagents, the rich chemistry available from such reactions also indicates that it is a fruitful area for investigation.[186]

Acknowledgment

Funding from the National Science Foundation (CBET-0828767) is gratefully acknowledged.

References

1. J. K. Norskov, T. Bligaard, J. Rossmeisl and C. H. Christensen, *Nature Chem.*, 2009, **1**, 37.
2. J. Greeley, I. E. L. Stephens, A. S. Bondarenko, T. P. Johansson, H. A. Hansen, T. F. Jaramillo, J. Rossmeisl, I. Chorkendorff and J. K. Norskov, *Nature Chem.*, 2009, **1**, 552.
3. P. Ferrin, A. U. Nilekar, J. Greeley, M. Mavrikakis and J. Rossmeisl, *Surf. Sci.*, 2008, **602**, 3424.
4. J. Greeley and M. Mavrikakis, *Catal. Today*, 2006, **111**, 52.
5. M. P. Andersson, T. Bligaard, A. Kustov, K. E. Larsen, J. Greeley, T. Johannessen, C. H. Christensen and J. K. Norskov, *J. Catal.*, 2006, **239**, 501.
6. A. Boisen, S. Dahl, J. K. Norskov and C. H. Christensen, *J. Catal.*, 2005, **230**, 309.
7. E. Nikolla, J. Schwank and S. Linic, *J. Catal.*, 2007, **250**, 85.
8. M. Neurock, *J. Catal.*, 2003, **216**, 73.
9. R. A. van Santen and H. Kuipers, *Adv. Catal.*, 1987, **35**, 265.
10. J. R. Monnier, in *3rd World Congress on Oxidation Catal.* 1997; Vol. 110, p. 135.
11. S. Linic and M. A. Barteau, *J. Am. Chem. Soc.*, 2004, **126**, 8086.
12. J. R. Monnier, J. L. Stavinoha and G. W. Hartley, *J. Catal.*, 2004, **226**, 321.
13. J. R. Monnier, J. L. Stavinoha and R. L. Minga, *J. Catal.*, 2004, **226**, 401.
14. S. Linic, J. Jankowiak and M. A. Barteau, *J. Catal.*, 2004, **224**, 489.
15. H. Piao, K. Adib, S. Linic, M. Enever, J. Hrbek and M. A. Barteau, *Abstracts Papers Am. Chem. Soc.*, 2003, **226**, U377.
16. S. Linic and M. A. Barteau, *J. Catal.*, 2003, **214**, 200.
17. S. Linic and M. A. Barteau, *J. Am. Chem. Soc.*, 2003, **125**, 4034.
18. J. W. Medlin, M. A. Barteau and J. M. Vohs, *J. Mol. Catal. A*, 2000, **163**, 129.

19. P. Christopher and S. Linic, *J. Am. Chem. Soc.*, 2008, **130**, 11264.
20. J. W. Medlin and M. A. Allendorf, *J. Phys. Chem. B*, 2003, **107**, 217.
21. D. H. Mei, M. Neurock and C. M. Smith, *J. Catal.*, 2009, **268**, 181.
22. D. Mei, P. A. Sheth, M. Neurock and C. M. Smith, *J. Catal.*, 2006, **242**, 1.
23. P. A. Sheth, M. Neurock and C. M. Smith, *J. Phys. Chem. B*, 2005, **109**, 12449.
24. P. A. Sheth, M. Neurock and C. M. Smith, *J. Phys. Chem. B*, 2003, **107**, 2009.
25. P. S. Cremer, X. C. Su, Y. R. Shen and G. A. Somorjai, *J. Phys. Chem. B*, 1997, **101**, 6474.
26. F. Studt, F. Abild-Pedersen, T. Bligaard, R. Z. Sorensen, C. H. Christensen and J. K. Norskov, *Science*, 2008, **320**, 1320.
27. Y. Kalinci, A. Hepbasli and I. Dincer, *Int. J. Hydrog. Energy*, 2009, **34**, 8799.
28. M. M. Yung, W. S. Jablonski and K. A. Magrini-Bair, *Energy Fuels*, 2009, **23**, 1874.
29. R. J. Farrauto and C. H. Bartholomew, *Fundamentals Industrial Catalytic Processes*, Blackie Academic and Professional, New York, 1997.
30. D. A. Laird, R. C. Brown, J. E. Amonette and J. Lehmann, *Biofuels Bioprod. Biorefining*, 2009, **3**, 547.
31. B. Digman, H. S. Joo and D. S. Kim, *Environ. Prog. Sustain. Energy*, 2009, **28**, 47.
32. D. Mohan, C. U. Pittman and P. H. Steele, *Energy Fuels*, 2006, **20**, 848.
33. A. V. Bridgwater, *Chem. Eng. J.*, 2003, **91**, 87.
34. Y. C. Lin, J. Cho, G. A. Tompsett, P. R. Westmoreland and G. W. Huber, *J. Phys. Chem. C*, 2009, **113**, 20097.
35. A. Demirbas, *Energy Conv. Manag.*, 2000, **41**, 633.
36. T. R. Carlson, G. A. Tompsett, W. C. Conner and G. W. Huber, *Top. Catal.*, 2009, **52**, 241.
37. C. E. Wyman, S. R. Decker, M. E. Himmel, J. W. Brady, C. E. Skopec and L. Viikari, in *Polysaccharides*, 2nd edn, ed. S. Dumitriu, Marcel Dekker, New York, 2005, p. 995.
38. A. Fukuoka and P. L. Dhepe, *Chem. Rec.*, 2009, **9**, 224.
39. F. Xu and H. S. Ding, *Appl. Catal. Gen.*, 2007, **317**, 70.
40. J. N. Chheda, G. W. Huber and J. A. Dumesic, *Angew. Chem., Int. Ed. Engl.*, 2007, **46**, 7164.
41. G. W. Huber and J. A. Dumesic, *Catal. Today*, 2006, **111**, 119.
42. G. W. Huber, J. N. Chheda, C. J. Barrett and J. A. Dumesic, *Science*, 2005, **308**, 1446.
43. G. W. Huber, R. D. Cortright and J. A. Dumesic, *Angew. Chem., Int. Ed. Engl.*, 2004, **43**, 1549.
44. T. Werpy and G. Petersen, *Top Value Added Chemicals from Biomass*, National Renewable Energy Laboratory and Pacific Northwest National Laboratory Report, 2004.
45. V. F. Wendisch, M. Bott and B. J. Eikmanns, *Curr. Opin Microbiol.*, 2006, **9**, 268.

46. J. Zaldivar, A. Martinez and L. O. Ingram, *Biotechnol. Bioeng.*, 1999, **65**, 24.
47. A. Knop-Gericke, E. Kleimenov, M. Havecker, R. Blume, D. Teschner, S. Zafeiratos, R. Schlogl, V. I. Bukhtiyarov, V. V. Kaichev, I. P. Prosvirin, A. I. Nizovskii, H. Bluhm, A. Barinov, P. Dudin and M. Kiskinova, *Adv. Catal.*, 2009, **52**, 213.
48. G. Ertl, *J. Mol. Catal. Chem.*, 2002, **182**, 5.
49. H. Over, Y. D. Kim, A. P. Seitsonen, S. Wendt, E. Lundgren, M. Schmid, P. Varga, A. Morgante and G. Ertl, *Science*, 2000, **287**, 1474.
50. P. A. Redhead, *Vacuum*, 1962, **12**, 203.
51. R. J. Madix, *CRC Critical Reviews in Solid State and Materials Sciences*, 1978, **7**, 143.
52. H. Ibach and D. L. Mills, *Electron Energy Loss Spectroscopy and Surface Vibrations*, Academic Press, Inc., New York, 1982.
53. D. P. Woodruff and T. A. Delchar, *Modern Techniques Surface Science*, Cambridge University Press, Cambridge, UK, 1986.
54. J. Stohr, *NEXAFS Spectroscopy*, Springer, New York, 1996, Vol. 25.
55. P. Gambardella, Z. Sljivancanin, B. Hammer, M. Blanc, K. Kuhnke and K. Kern, *Phys. Rev. Lett.*, 2001, **87**, 056103: 1.
56. J. W. Medlin and M. A. Barteau, *J. Phys. Chem. B*, 2001, **105**, 10054.
57. J. W. Medlin, A. B. Sherrill, J. G. G. Chen and M. A. Barteau, *J. Phys. Chem. B*, 2001, **105**, 3769.
58. R. T. Vang, K. Honkala, S. Dahl, E. K. Vestergaard, J. Schnadt, E. Laegsgaard, B. S. Clausen, J. K. Norskov and F. Besenbacher, *Surf. Sci.*, 2006, **600**, 66.
59. P. S. Cremer and G. A. Somorjai, *J. Chem. Soc., Faraday Trans.*, 1995, **91**, 3671.
60. J. E. Demuth, *Surf. Sci.*, 1979, **84**, 315.
61. M. Salmeron and G. A. Somorjai, *J. Phys. Chem.*, 1982, **86**, 341.
62. R. J. Koestner, J. Stohr, J. L. Gland and J. A. Horsley, *Chem. Phys. Lett.*, 1984, **105**, 332.
63. B. E. Bent, C. M. Mate, C. T. Kao, A. J. Slavin and G. A. Somorjai, *J. Phys. Chem.*, 1988, **92**, 4720.
64. P. S. Cremer, X. C. Su, Y. R. Shen and G. A. Somorjai, *J. Am. Chem. Soc.*, 1996, **118**, 2942.
65. M. Neurock, V. Pallassana and R. A. van Santen, *J. Am. Chem. Soc.*, 2000, **122**, 1150.
66. W. Erley, A. M. Baro and H. Ibach, *Surf. Sci.*, 1982, **120**, 273.
67. J. L. Solomon, R. J. Madix and J. Stohr, *J. Chem. Phys.*, 1990, **93**, 8379.
68. C. J. Jenks, B. E. Bent, N. Bernstein and F. Zaera, *Surf. Sci.*, 1992, **277**, L89.
69. M. Mavrikakis and M. A. Barteau, *J. Mol. Catal. Chem.*, 1998, **131**, 135.
70. X. Y. Liu, R. J. Madix and C. M. Friend, *Chem. Soc. Rev.*, 2008, **37**, 2243.
71. N. R. Avery, *Surf. Sci.*, 1983, **125**, 771.
72. J. L. Davis and M. A. Barteau, *Surf. Sci.*, 1987, **187**, 387.
73. J. B. Benziger and R. J. Madix, *J. Catal.*, 1980, **65**, 36.

74. P. H. McBreen, W. Erley and H. Ibach, *Surf. Sci.*, 1983, **133**, L469.

75. J. Hrbek, R. A. Depaola and F. M. Hoffmann, *J. Chem. Phys.*, 1984, **81**, 2818.

76. P. Gazdzicki and P. Jakob, *J. Phys. Chem. C*, 2010, **114**, 2655.

77. J. E. Demuth and H. Ibach, *Chem. Phys. Lett.*, 1979, **60**, 395.

78. S. Johnson and R. J. Madix, *Surf. Sci.*, 1981, **103**, 361.

79. K. Christmann and J. E. Demuth, *J. Chem. Phys.*, 1982, **76**, 6308.

80. K. Christmann and J. E. Demuth, *J. Chem. Phys.*, 1982, **76**, 6318.

81. F. Solymosi, A. Berko and T. I. Tarnoczi, *Surf. Sci.*, 1984, **141**, 533.

82. S. R. Bare, J. A. Stroscio and W. Ho, *Surf. Sci.*, 1985, **150**, 399.

83. A. K. Bhattacharya, M. A. Chesters, M. E. Pemble and N. Sheppard, *Surf. Sci.*, 1988, **206**, L845.

84. C. Houtman and M. A. Barteau, *Langmuir*, 1990, **6**, 1558.

85. J. E. Parmeter, X. D. Jiang and D. W. Goodman, *Surf. Sci.*, 1990, **240**, 85.

86. N. Kizhakevariam and E. M. Stuve, *Surf. Sci.*, 1993, **286**, 246.

87. R. B. Barros, A. R. Garcia and L. M. Ilharco, *J. Phys. Chem. B*, 2001, **105**, 11186.

88. G. C. Wang, Y. H. Zhou and J. Nakamura, *J. Chem. Phys.*, 2005, **122**.

89. J. Greeley and M. Mavrikakis, *J. Catal.*, 2002, **208**, 291.

90. I. E. Wachs and R. J. Madix, *Surf. Sci.*, 1978, **76**, 531.

91. M. Bowker and R. J. Madix, *Surf. Sci.*, 1980, **95**, 190.

92. A. F. Lee, D. E. Gawthrope, N. J. Hart and K. Wilson, *Surf. Sci.*, 2004, **548**, 200.

93. S. M. Gates, J. N. Russell and J. T. Yates, *Surf. Sci.*, 1986, **171**, 111.

94. J. L. Davis and M. A. Barteau, *Surf. Sci.*, 1990, **235**, 235.

95. N. Kruse, M. Rebholz, V. Matolin, G. K. Chuah and J. H. Block, *Surf. Sci.*, 1990, **238**, L457.

96. M. Rebholz and N. Kruse, *J. Chem. Phys.*, 1991, **95**, 7745.

97. F. Solymosi, A. Berko and Z. Toth, *Surf. Sci.*, 1993, **285**, 197.

98. C. J. Houtman and M. A. Barteau, *J. Catal.*, 1991, **130**, 528.

99. G. S. Jones, M. Mavrikakis, M. A. Barteau and J. M. Vohs, *J. Am. Chem. Soc.*, 1998, **120**, 3196.

100. L. E. Murillo and J. G. Chen, *Surf. Sci.*, 2008, **602**, 2412.

101. N. F. Brown and M. A. Barteau, *J. Phys. Chem.*, 1996, **100**, 2269.

102. N. F. Brown and M. A. Barteau, *Langmuir*, 1992, **8**, 862.

103. M. A. Henderson, P. L. Radloff, J. M. White and C. A. Mims, *J. Phys. Chem.*, 1988, **92**, 4111.

104. R. W. McCabe, C. L. Dimaggio and R. J. Madix, *J. Phys. Chem.*, 1985, **89**, 854.

105. M. Bowker and R. J. Madix, *Surf. Sci.*, 1981, **102**, 542.

106. C. Houtman and M. A. Barteau, *J. Phys. Chem.*, 1991, **95**, 3755.

107. F. Delbecq and F. Vigne, *J. Phys. Chem. B*, 2005, **109**, 10797.

108. S. Linic and M. A. Barteau, *J. Am. Chem. Soc.*, 2002, **124**, 310.

109. J. Kim, H. B. Zhao, C. Panja, A. Olivas and B. E. Koel, *Surf. Sci.*, 2004, **564**, 53.

110. D. Stacchiola, G. Wu, M. Kaltchev and W. T. Tysoe, *Surf. Sci.*, 2001, **486**, 9.
111. C. Benndorf, B. Nieber and B. Kruger, *Surf. Sci.*, 1987, **189**, 511.
112. R. Shekhar, M. A. Barteau, P. V. Plank and J. M. Vohs, *Surf. Sci.*, 1997, **384**, L815.
113. R. Shekhar, M. A. Barteau and J. Vacuum Science & Technology, *Vacuum Surfaces and Films*, 1996, **14**, 1469.
114. N. F. Brown and M. A. Barteau, *Surf. Sci.*, 1993, **298**, 6.
115. F. Solymosi, J. Cserenyi and L. Ovari, *Catal. Lett.*, 1997, **44**, 89.
116. F. Solymosi and G. Klivenyi, *Surf. Sci.*, 1998, **409**, 241.
117. C. Houtman and M. A. Barteau, *Surf. Sci.*, 1991, **248**, 57.
118. B. H. Toby, N. R. Avery, A. B. Anton and H. Weinberg, *J. Electron Spectroscopy and Related Phenomena*, 1983, **29**, 317.
119. N. R. Avery, *Appl. Surf. Sci.*, 1982, **11–2**, 774.
120. M. A. Barteau, M. Bowker and R. J. Madix, *Surf. Sci.*, 1980, **94**, 303.
121. J. L. Falconer and R. J. Madix, *Surf. Sci.*, 1974, **46**, 473.
122. M. Ito and W. Suetaka, *J. Catal.*, 1978, **54**, 13.
123. R. J. Madix and S. G. Telford, *Surf. Sci.*, 1992, **277**, 246.
124. F. Solymosi and I. Kovacs, *Surf. Sci.*, 1991, **259**, 95.
125. F. Solymosi, J. Kiss and I. Kovacs, *J. Vacuum Sci. & Technology: Vacuum Surfaces and Films*, 1987, **5**, 1108.
126. N. R. Avery, B. H. Toby, A. B. Anton and W. H. Weinberg, *Surf. Sci.*, 1982, **122**, L574.
127. L. E. Firment and G. A. Somorjai, *J. Vacuum Sci. & Technology*, 1980, **17**, 574.
128. Q. Y. Gao and J. C. Hemminger, *J. Electron Spectroscopy and Related Phenomena*, 1990, **54**, 667.
129. G. R. Schoofs and J. B. Benziger, *Surf. Sci.*, 1984, **143**, 359.
130. C. M. Horiuchi and J. W. Medlin, *Surf. Sci.*, 2010, **604**, 98.
131. L. Xu and Y. Xu, *Surf. Sci.*, 2010, **604**, 887.
132. J. C. de Jesus and F. Zaera, *J. Molecular Catal. Chem.*, 1999, **138**, 237.
133. J. C. de Jesus and F. Zaera, *Surf. Sci.*, 1999, **430**, 99.
134. F. Delbecq and P. Sautet, *J. Catal.*, 1995, **152**, 217.
135. R. Shekhar and M. A. Barteau, *Surf. Sci.*, 1994, **319**, 298.
136. J. L. Davis and M. A. Barteau, *J. Mol. Catal.*, 1992, **77**, 109.
137. S. T. Marshall, C. M. Horiuchi, W. Y. Zhang and J. W. Medlin, *J. Phys. Chem. C*, 2008, **112**, 20406.
138. N. F. Brown and M. A. Barteau, *J. Am. Chem. Soc.*, 1992, **114**, 4258.
139. F. Delbecq and P. Sautet, *J. Catal.*, 2002, **211**, 398.
140. J. Haubrich, D. Loffreda, F. Delbecq, P. Sautet, A. Krupski, C. Becker and K. Wandeltt, *J. Phys. Chem. C*, 2009, **113**, 13947.
141. J. Haubrich, D. Loffreda, F. Delbecq, P. Sautet, Y. Jugnet, A. Krupski, C. Becker and K. Wandelt, *J. Phys. Chem. C*, 2008, **112**, 3701.
142. D. Loffreda, F. Delbecq, F. Vigne and P. Sautet, *J. Am. Chem. Soc.*, 2006, **128**, 1316.
143. J. Haubrich, D. Loffreda, F. Delbecq, Y. Jugnet, P. Sautet, A. Krupski, C. Becker and K. Wandelt, *Chem. Phys. Lett.*, 2006, **433**, 188.

144. D. Loffreda, F. Delbecq and P. Sautet, *Chem. Phys. Lett.*, 2005, **405**, 434.
145. D. Loffreda, Y. Jugnet, F. Delbecq, J. C. Bertolini and P. Sautet, *J. Phys. Chem. B*, 2004, **108**, 9085.
146. R. Hirschl, F. Delbecq, P. Sautet and J. Hafner, *J. Catal.*, 2003, **217**, 354.
147. F. Delbecq and P. Sautet, *J. Catal.*, 2003, **220**, 115.
148. A. S. Loh, S. W. Davis and J. W. Medlin, *J. Am. Chem. Soc.*, 2008, **130**, 5507.
149. M. T. Schaal, M. P. Hyman, M. Rangan, S. Ma, C. T. Williams, J. R. Monnier and J. W. Medlin, *Surf. Sci.*, 2009, **603**, 690.
150. M. J. Knight, F. Allegretti, E. A. Kroger, M. Polcik, C. L. A. Lamont and D. P. Woodruff, *Surf. Sci.*, 2008, **602**, 2524.
151. A. Loui and S. Chiang, *Appl. Surf. Sci.*, 2004, **237**, 559.
152. T. E. Caldwell and D. P. Land, *J. Phys. Chem. B*, 1999, **103**, 7869.
153. R. M. Ormerod, C. J. Baddeley, C. Hardacre and R. M. Lambert, *Surf. Sci.*, 1996, **360**, 1.
154. T. E. Caldwell, I. M. Abdelrehim and D. P. Land, *J. Am. Chem. Soc.*, 1996, **118**, 907.
155. J. W. Medlin, C. M. Horiuchi and M. Rangan, *Top. Catal.*, 2010, **53**, 1179.
156. J. L. Solomon and R. J. Madix, *J. Phys. Chem.*, 1987, **91**, 6241.
157. R. L. Brainard, C. G. Peterson and R. J. Madix, *J. Am. Chem. Soc.*, 1989, **111**, 4553.
158. F. Calaza, D. Stacchiola, M. Neurock and W. T. Tysoe, *Surf. Sci.*, 2005, **598**, 263.
159. C. M. Horiuchi, B. T. Israel and J. W. Medlin, *J. Phys. Chem. C*, 2009, **113**, 14900.
160. A. J. Capote and R. J. Madix, *J. Am. Chem. Soc.*, 1989, **111**, 3570.
161. C. R. Ayre and R. J. Madix, *Surf. Sci.*, 1994, **303**, 279.
162. C. R. Ayre and R. J. Madix, *Surf. Sci.*, 1994, **303**, 297.
163. O. Skoplyak, M. A. Barteau and J. G. G. Chen, *J. Phys. Chem. B*, 2006, **110**, 1686.
164. O. Skoplyak, M. A. Barteau and J. G. G. Chen, *Surf. Sci.*, 2008, **602**, 3578.
165. O. Skoplyak, C. A. Menning, M. A. Barteau and J. G. G. Chen, *Top. Catal.*, 2008, **51**, 49.
166. A. L. Stottlemyer, H. Ren and J. G. Chen, *Surf. Sci.*, 2009, **603**, 2630.
167. O. Skoplyak, M. A. Barteau and J. G. G. Chen, *Catal. Today*, 2009, **147**, 150.
168. M. M. M. Jansen, B. E. Nieuwenhuys and H. Niemantsverdriet, *Chem. Sus. Chem.*, 2009, **2**, 883.
169. M. Bowker and R. J. Madix, *Surf. Sci.*, 1982, **116**, 549.
170. K. T. Queeney, C. R. Arumainayagam, M. K. Weldon, C. M. Friend and M. Q. Blumberg, *J. Am. Chem. Soc.*, 1996, **118**, 3896.
171. G. W. Huber, J. W. Shabaker and J. A. Dumesic, *Science*, 2003, **300**, 2075.
172. C. L. Bianchi, P. Canton, N. Dimitratos, F. Porta and L. Prati, *Catal. Today*, 2005, **102**, 203.

173. N. Dimitratos, A. Villa and L. Prati, *Catal. Lett.*, 2009, **133**, 334.
174. C. H. C. Zhou, J. N. Beltramini, Y. X. Fan and G. Q. M. Lu, *Chem. Soc. Rev.*, 2008, **37**, 527.
175. A. Behr, J. Eilting, K. Irawadi, J. Leschinski and F. Lindner, *Green Chem.*, 2008, **10**, 13.
176. A. Corma, S. Iborra and A. Velty, *Chem. Rev.*, 2007, **107**, 2411.
177. C. Delhomme, D. Weuster-Botz and F. E. Kuhn, *Green Chem.*, 2009, **11**, 13.
178. R. M. Deshpande, V. V. Buwa, C. V. Rode, R. V. Chaudhari and P. L. Mills, *Catal. Commun.*, 2002, **3**, 269.
179. Y. Roman-Leshkov and J. A. Dumesic, *Topics Catal.*, 2009, **52**, 297.
180. R. M. West, Z. Y. Liu, M. Peter, C. A. Gartner and J. A. Dumesic, *J. Mol. Catal. Chem.*, 2008, **296**, 18.
181. J. N. Chheda and J. A. Dumesic, *Catal. Today*, 2007, **123**, 59.
182. S. Crossley, J. Faria, M. Shen and D. E. Resasco, *Science*, 2010, **327**, 68.
183. V. Schiavo, G. Descotes and J. Mentech, *Bull. Soc. Chim. Fr.*, 1991, **704**.
184. D. M. Meier, A. Urakawa and A. Baiker, *J. Phys. Chem. C*, 2009, **113**, 21849.
185. S. P. Devarajan, J. A. Hinojosa and J. F. Weaver, *Surf. Sci.*, 2008, **602**, 3116.
186. C. M. Horiuchi and J. W. Medlin, *Langmuir*, 2010, **26**, 13320.

Dilute Acid and Hydrothermal Pretreatment of Cellulosic Biomass

DEEPTI TANJORE,[1] JIAN SHI[1] AND CHARLES E. WYMAN[1,2]

[1] Center for Environmental Research and Technology and; [2] Chemical and Environmental Engineering Department, University of California, Riverside, California, 92507

4.1 Introduction

The natural recalcitrance of lignocellulosic biomass, due its complex structural matrix, resists release of sugars,[1] and pretreatment is critical for opening up the complex plant cell wall structure for fractionation into its components and high yields vital for economic success.[2–4] Hydrothermal and dilute acid pretreatment have proven to be relatively simple and effective approaches to reducing this natural recalcitrance while also releasing sugars during these operations.[5]

Water in the form of steam provided mechanical power for industry since the eighteenth century. In the 1930s, steam was used for production of fiberboard from wood at high temperatures (220–270 °C) and very short residence times (4–90 s) followed by rapid release in the so-called Masonite steam explosion process.[6] More recently, similar technology was applied to produce feed for ruminants[7] and pretreat lignocellulosic biomass.[8] Lignocellulosic biomass was heated directly in liquid water or by percolation of hot water through biomass

RSC Energy and Environment Series No. 4
Chemical and Biochemical Catalysis for Next Generation Biofuels
Edited by Blake Simmons

Published by the Royal Society of Chemistry, www.rsc.org

to remove hemicellulose.[9,10] We designate these approaches as hydrothermal pretreatment to distinguish them from processes that use acid or other catalysts. A number of other names have also been used over the years including water-only treatment, autohydrolysis,[11] hydrothermolysis,[9,12] aqueous or steam/aqueous fractionation,[13,14] uncatalyzed solvolysis,[15] and aquasolv[16] to describe the application of water in pretreating biomass.

Dilute acid has been employed for some time as a catalyst to hydrolyze woody materials to sugars, although initial emphasis was on releasing glucose from cellulose. The first commercially significant wood hydrolysis process, the Scholler process, was developed in Germany in 1931.[17] Wood waste was preheated with live steam and treated with 0.5% sulfuric acid for 45 minutes at 130 °C to release sugars which were fermented to ethanol or glycerol or used as a feed to yeast inocula.[17] However, this approach suffered from low yields of sugars and considerable plant downtime due to fouling by lignin and products formed by hemicellulose degradation.[17] An improved version of the Scholler process, the Madison process, was developed in the USA,[18] with dilute sulfuric acid (0.5–0.6%) passed through a biomass bed at high temperatures (150–180 °C) but with less reaction time to avoid degradation of dissolved sugars. Nonetheless, despite higher temperatures and shorter reaction times in the Madison process, total reducing sugar recovery was limited to about 49% in bark.

Because sugars are released more readily from hemicellulose than from cellulose and xylose degrades rapidly at the time/temperature combinations required to breakdown cellulose, hemicellulose sugar yields were very low at conditions that realized high glucose yields from the Madison, Scholler, and other one-step processes. Therefore, a two-step approach was developed by Dunning and Lathrop at USDA Peoria.[19] In the first step, dilute sulfuric acid (4.9–9.8%) was used at 98 °C for a reaction time of between 50 and 185 minutes to primarily release sugars from the hemicellulose fraction of biomass, and the liquid stream, containing primarily the pentoses from hemicellulose, was separated from the remaining solids after the reaction. Then, the acid-treated solids were re-treated with 8% sulfuric acid at 120–130 °C for 7–10 minutes to obtain glucose from cellulose. Acids other than sulfuric acid, including hydrochloric (HCl), phosphoric (H_3PO_4), and hydrofluoric acids (HF), were also evaluated for their abilities to improve sugar yields.[20,21] However, the recovery and re-use of these more expensive acids were not economically viable, tending to override any advantages in sugar recovery.[20] Accordingly, dilute sulfuric acid and possibly sulfur dioxide tend to be favored because of their low cost and more desirable safety and environmental aspects.

Although early processes sought to apply acid for cellulose breakdown to glucose, the reaction kinetics limited yields and were not economically attractive, leading to use of enzymes for cellulose conversion during the 1980s to take advantage of the technology improvements in modern biotechnology and obtain higher sugar yields. However, because pretreatment was needed to realize high yields, Hsu *et al.*[22] proposed applying dilute acid to treat lignocellulosic biomass prior to enzymatic hydrolysis rather than using it for conversion of both hemicellulose and cellulose. In this case, it was found that most of the hemicellulose in biomass could be hydrolyzed to dissolved sugars with 0.1–1.0% sulfuric acid over a range of 95–120 °C[22] while the remaining solids primarily contained cellulose, with the biomass altered enough to make the cellulose accessible to enzymes attack. It was further shown that enzymatic hydrolysis could release no more than about 40% of the glucose from untreated biomass, while yields could be about 90% following pretreatment with acid.[22] Pretreatment of lignocellulosic substrates with dilute acid and hot water was patented in 1980 by Grethlein at Dartmouth College.[23] In this case, biomass slurry was maintained between 180 and 230 °C in the presence of low sulfuric acid concentration (about 1%) for a small time period, usually less than 1 minute. As an alternative approach, hydrothermal pretreatment can be performed with temperatures between 180 and 300 °C and reaction times of between 5 and 1 minute, respectively. Dilute acid not only solubilized most of the hemicellulose as monomeric sugars but also enhanced cellulose digestion during enzymatic hydrolysis.[24] Since then, a number of combinations of process conditions and reactor designs have been tested to improve dilute acid pretreatment, and hemicellulose reaction chemistry and kinetics, the relationship between pretreatment conditions and enzymatic digestion of pretreated biomass, the effects of feedstock variability, and the techno-economics of hydrothermal and particularly dilute acid pretreatment. These aspects are summarized in the following sections.

4.2   Pretreatment Chemistry

Dilute acid and hydrothermal pretreatments hydrolyze much of the hemicellulose plus a small amount of the cellulose to release their constituent sugars and make the remaining solids available for enzymatic hydrolysis with high yields. Hydrolysis of a molecule involves its cleavage into two parts by addition of water molecules (H_2O) that split into hydrogen cations (H^+, conventionally referred to as protons) and hydroxide anions (OH^-). Each of the ions attaches to a different part of the parent molecule, inducing H-bond changes and breaking it into two parts. For cellulose, the reaction stoichiometry can be represented as:

$$nC_6H_{10}O_5 + nH_2O \rightarrow nC_6H_{12}O_6 \tag{1}$$

For the reaction of xylan in hemicellulose to form xylose, the overall chemistry is:

$$nC_5H_8O_4 + nH_2O \rightarrow nC_5H_{10}O_5 \tag{2}$$

Hydrothermal pretreatment, a process traditionally often referred to as autohydrolysis or thermohydrolysis in the literature, involves hydrolysis of lignocellulosic biomass in the presence of 'free water' during pretreatment. The term autohydrolysis was based on the hypothesis that acidic compounds (mainly acetic acid) released from acetylated hemicellulose catalyze hydrolysis as evidenced by the positive correlation of hemicellulose hydrolysis with accumulation of acetic acid during hydrothermal pretreatment.[11] However, this assumption is not completely validated, as near complete xylan removal is observed in flowthrough reactors, where acetic acid is virtually absent due to the continuous flow of water.[25] In addition, data from Datar *et al.*[26] suggests that a significant portion of the acetic acid is released late in hydrolysis.

Addition of small amounts of mineral acids, usually H_2SO_4, improves hydrolysis of hemicellulose significantly at reduced temperatures compared with hydrothermal pretreatment by increasing the concentration of protons. For example, nearly complete hemicellulose removal was achieved when corn stover was treated with 0.5–1.0 wt% H_2SO_4 solution and incubated at 140 °C for 30 minutes, or at 160 °C for as little as 5–10 minutes.[27] Alternatively, 1–5% sulfur dioxide (SO_2) gas can be added to moist biomass chips and heated to 150–200 °C for 3–20 minutes to hydrolyze hemicellulose.[28,29] Sulfur dioxide rapidly diffuses into biomass pores before it is converted to H_2SO_3, providing performance advantages compared to direct use of an acid catalyst. Steam pretreatment of hardwood or softwood using SO_2 can hydrolyze more than 80% of the hemicellulose to sugar monomers and is far less corrosive than mineral acids.[30]

In addition to hydrolysis reactions, mechanical forces during the explosive expansion of steam were believed to open up the plant material to expose fibers, presumably increasing the accessibility of polysaccharides for subsequent enzymatic hydrolysis. However, evidence to support this hypothesis only existed for steam explosion that involves saturation of plant material with steam followed by rapid decompression (pressure release). More recent research suggested that chemical changes occurring prior to rapid decompression were more important than mechanical disruption.[30] These chemical changes include the aforementioned hemicellulose hydrolysis and concurrent condensation and relocation of lignin. Softening of the middle lamella components during stream treatment led to structural changes and migration of lignin.[31] Softening of lignin and hemicelluloses were said to take place at temperatures as low as 100 °C or less in the presence of water, although the anhydrous glass transition of pure hemicellulose and lignin polymers need much higher temperatures.[32] At higher temperatures, lignin in the pulp residue was observed to coalesce into small spherical globules of size less than 5–10 μm diameter.[31] Further recent visual evidence using SEM/TEM techniques reported the coalescence and migration of lignin during dilute acid pretreatment at temperatures above the lignin phase transition range.[33] These results also indicated that lignin decompartmentalization and relocalization is likely to be an important contributor of the improved digestibility of lignocellulosic biomass, a view supported by observations reported for flowthrough pretreatments.[34]

Overend and Chornet[13] summarized the chemical consequences of hydrothermal pretreatment of lignocellulosics as the rupture of glycosidic bonds and disaggregation of lignin–carbohydrate complexes primarily due to a hydrolytic mechanism.[13] Additional effects were said to be extensive cleavage of C–O–C bonds of lignin[35,36] and lignin depolymerization down to relatively small subunits of around 2000–3000 mass-average molecular weight values.[37] At very high treatment severities, crystallinity of the residual cellulose increased due to pyrolytic dehydrations.[38] Dilute acid and hydrothermal pretreatment of lignocellulosic biomass was thought to also impact physical as well as chemical changes in both the ultra-structure and functionalities of constitute polymers.[39]

The result of hydrothermal and dilute acid pretreatment is a cellulose-enriched solid stream and a liquid stream containing mainly soluble mono and polymeric pentose sugars, and smaller quantities of hexose sugars, acetic acid, carboxylic acids, furfural from dehydration of arabinose and xylose, hydroxymethylfurfural (HMF) from degradation of glucose, and low-molecular weight phenols and aldehydes derived from lignin. Many of these and other components released during or formed by hydrothermal and dilute acid pretreatment have been identified as potential inhibitors to downstream enzymatic hydrolysis and fermentation steps.[40,41]

For example, acetic acid inhibits fermentations beyond a threshold value as low as 0.5 g/L for many ethanol-fermenting microorganisms. A few microbial strains with enhanced acetate tolerances have been developed to counter the inhibition effect.[42–45] In addition; furfural, HMF, and phenolic lignin derivatives are reported to be detrimental to cell growth and ethanol fermentation at even lower concentrations.[46–48]

Several methods have been developed to remove inhibitors from pretreatment hydrolyzate streams and thereby improve fermentation efficiency including overliming, steam stripping, ion-exchange, solvent and reactive membrane extraction, treatment with activated carbon and zeolites, and organism acclimation.[42,49–51] Although overliming using calcium hydroxide (lime) is relatively inexpensive, some of the sugars degrade during this step, and the $CaSO_4$ (gypsum) formed during treatment presents a waste disposal problem.[52] Gypsum also has reverse solubility characteristics that will foul downstream heat exchangers unless proper countermeasures are taken.[53] Maximizing production of sugars from the coupled operations of pretreatment and enzymatic hydrolysis is typically the most important consideration in designing pretreatment processes.

4.3 Laboratory Reactors

Various reactor types have been employed to define optimal conditions for pretreatment. High hemicellulose sugar yields, generally measured in terms of xylan dissolved in the liquid, and improved digestibility of the cellulose and, more recently, the other carbohydrates in the pretreated solids, have been considered as benchmarks for successful pretreatment.[3,4] Other than switching between batch and continuous systems, several single and combinations of reactor have been tested, some of which will be illustrated below. Key requirements include the ability to differentiate reaction kinetics from heat, mass, or momentum transfer effects to assure that the data can be interpreted and applied accurately.

4.3.1 Batch Reactors

Mostly used for the study of kinetics of acid hydrolysis, batch reactors are the most commonly found laboratory-scale pretreatment reactors. In general, sugar yields increase initially with temperature and/or reaction times in batch reactors but then drop off with extended time due to dehydration of the sugars released to furfural and other degradation products that are also inhibitory to fermentation.[54] The capacities of laboratory batch reactors can vary from about 0.2 mg for high throughput 96-well plates employed for screening[55] to 400 grams or more for steam-explosion reactors.[56,57] Some examples are given below.

4.3.1.1 Tube Reactors

Reactors made from tubing provide the simplest and least expensive way to pretreat biomass at the laboratory scale, as shown in Figure 4.1. However, it is critical that the reactor diameter be as small as possible to assure uniform heat transfer. An outside diameter of about 12.5 mm and a wall thickness of 0.8255 mm provides a practical limitation to accommodate loading and processing adequate quantities of biomass in the reactor.[58,59] Although the reactor length can be flexible as long as the entire device can be fully immersed in a sand bath or other heating device, 10 to 15 cm long tubes are typical for along with

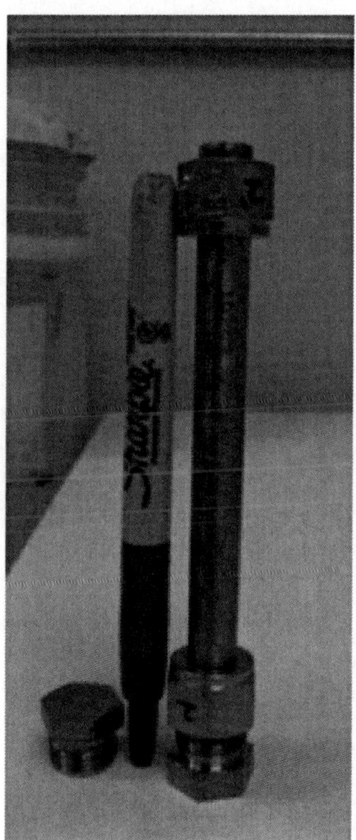

Figure 4.1 Tubular reactor made of tubing and capped by standard fittings.

several reactors typically processed at the same time. These reactors are made of appropriate materials such as Hastelloy C in the case of dilute acid pre-treatment, but may be closed with 304 stainless steel end caps outfitted with Teflon inserts if Hastelloy caps are not available.[60] Usually, about 1 g of dry biomass is loaded into each tube along with enough water or dilute acid to reach a 10% solids loading that facilitates pouring of the solids and liquid into the tube separately with a funnel. It is vital to leave enough head space in the reactor to allow water expansion at the higher reaction temperatures based on differences in specific volume of water over the range of temperatures spanned.[61]

4.3.1.2   Stirred Reactor

Stirred reactors such as made by Parr Instruments (Moline, IL) and shown in Figure 4.2, are commonly used for pretreating larger amounts of biomass with dilute acid, just water, or other chemicals such as ammonia, which are not possible to be tested in the tubes. Reactors such as these must be able to handle high pressures, typically have a volume of about 1 liter, and are made of appropriate materials of construction such as Hastelloy C, when used for dilute acid pretreatment. Solids loadings are limited to less than about 10% by weight to assure adequate mixing,[62] and the reactor may be outfitted with a turbine blade mixer for lower solids concentrations or a helical impeller to handle more

Figure 4.2   1-liter high pressure Parr reactor suspended by a chain hoist over a fluidized sand bath used for rapid heat up.

viscous biomass suspensions.[27] A thermocouple is inserted into the reactor through an appropriate opening to follow the inside temperature history. Although manufacturers generally supply an electrical jacket for heating the vessel, heating with a fluidized sand bath avoids prolonged heat up times of electrical jackets, thereby providing a more controlled thermal history.[63]

4.3.1.3   Steam Gun

Steam guns allow pretreatment of larger amounts of biomass than possible with either tubes or stirred reactors and also accommodate much higher solids concentrations. A steam gun is a cylindrical vessel with a port at the top for adding biomass initially, and equipped with a way to introduce steam directly into the vessel from the bottom to fluidize the contents and rapidly heat them to reaction conditions, and a valved opening at the bottom to allow discharge of the contents at the end of the reaction time and rapid cooling to stop the reaction as the pressure drops to atmospheric pressure. The vessel is typically made of an appropriate material to withstand the reaction temperature, pressure, and chemical environment. To minimize steam condensation beyond that needed to heat the reactor contents to the target temperature, the steam gun is jacketed for heating the sides with steam or wrapped in heating tape. For the example, as shown in Figure 4.3, the reactor has about a 4-inch (approx. 10 cm)

Figure 4.3   Steam gun reactor with biomass addition at the top and steam injection and reactor contents discharge at the bottom.

flanged opening at the top for loading biomass, while a 2-inch (approx. 5 cm) opening at the bottom is employed to release biomass after pretreatment. Before being fed into the reactor, biomass is presoaked in water or dilute acid and pressed to remove excess moisture to achieve the desired moisture level, typically about 50%. Thermocouples inserted through ports in the reactor allow the operator to follow the temperature history over the reaction period, with control by adjusting the steam input pressure.[57] Advantages of using a steam gun over other batch reactors include the ability to accurately control the temperature history for large quantities of biomass and to handle high solids concentrations ranging from 20 to 100% for the feed that result in higher sugar concentrations in the liquid hydrolyzate. Because solids concentrations of 50% or higher are favored for most projections of commercial designs and steam is generally chosen for such applications,[53–54,64] steam guns can provide data that is representative of that expected from commercial practice.

4.3.1.4 High Throughput Pretreatment and Hydrolysis (HTPH) Plates

Considerable time is required to pretreat biomass and analyze sugar release for all of the above methods, and none is able to pretreat very small amounts of biomass. Recently, a novel system was developed employing a 96-well plate format to rapidly screen about 0.2–0.5 mg of dry biomass in each well. Such a system allows the application of multiple combinations of biomass types, pretreatment conditions, and enzyme formulations.[55] As shown in Figure 4.4, the wells in the plates are of the same size as a standard 0.3 ml COSTAR 96 round bottom but are made of appropriate metal to withstand high pressures and chemicals for pretreatment. Unlike tubes, stirred pressure reactors, or

Figure 4.4   HTPH reactors automated with a robot to rapidly screen small amounts of biomass and evaluate multiple combinations of feedstock samples, pretreatment conditions, and enzyme formulations.

steam guns that allow complete removal of liquid hydrolyzate from the pretreated solids prior to enzymatic hydrolysis, pretreatment and subsequent enzymatic hydrolysis are performed in the same wells for the HTPH system to avoid difficulties with the time consuming process of solids handling between the two steps. Solids concentrations of 1–2% are employed to reduce inhibition of the enzymatic hydrolysis operation by sugars, oligomers, and other components released in pretreatment. HTPH systems perform similarly to conventional approaches at high enzyme loadings, and a novel steam chamber was devised to rapidly and uniformly heat and cool the wells.[55]

4.3.2 Continuous Reactors

Continuous reactors can achieve small reaction times, as low as a few seconds, and some configurations can minimize dehydration and degradation reactions of the sugars released from biomass by rapid removal of products. They can be categorized based on the flow of aqueous phase relative to the solid biomass, with leading options explained below.

4.3.2.1 Plug Flow Reactor

In a plug flow reactor (PFR), biomass moves with the liquid phase through the reactor, which is maintained at very high temperatures. Shorter reaction times are better controlled in PFR avoiding the degradation of sugar released from the biomass. About 10% higher sugar yields can be observed in PFR compared to batch reactor for the same feed due to continuous removal of the formed sugars. However, basic PFR designs can suffer from heat transfer limitations and accumulation of tars in industry-scale vessels.[2,55,65] In addition, a screw or other conveying means is needed to assure good residence time control and uniformity.[66]

4.3.2.2 Percolation/Flowthrough Reactor

For pretreatment applications, a percolation or flowthrough reactor is packed with biomass through which liquid is passed at high temperatures. Therefore, sugar produced in the reactor leaves in liquid phase, minimizing the opportunity for degradation. High flow rates in this reactor can give nearly theoretical sugar yields and high digestibility of the solids but lead to low sugar concentrations in the liquid stream.[25,34] Several variations in feed and temperature settings of the reactor were considered to obtain optimal sugar yields and concentrations.[54,67] The oldest examples of this type of reactor are the Scholler and Madison process, where several batches of liquid were replaced for one batch of biomass, although these devices focused on cellulose and not hemicellulose hydrolysis.[18] Percolation reactors rely on gravity to push water through the solids, while the term flowthrough reactor is applied when the liquid is pushed through the solids by applying high pressure. Thus, the former could be applied commercially while the latter is unlikely to be easily implemented.

4.3.2.3 *Countercurrent Reactor*

Countercurrent or moving bed reactors have liquid and biomass moving in opposite directions. One such configuration, called the Progressing Batch Reactor, relies on passing liquid through a number of percolation reactors connected in series with periodic addition of a new reactor along with a simultaneous removal of the oldest reactor from the chain. Liquid enters at the oldest reactor and leaves from the most recent one brought on line.[68] This configuration thus simulates countercurrent movement of solids and liquid while avoiding the challenges of pumping solids. An alternative countercurrent mode known as the Shrinking Bed Reactor is based on maintaining pressure on the solids with a spring or other arrangement so the void volume is minimized, thereby reducing the time for the sugars in the flowing liquid to degrade.[69] These reactors control reaction times well and can realize nearly theoretical yields for laboratory use but would be challenging to implement commercially.[54]

4.4 Reaction Kinetics and Severity Factor

Knowledge of reaction kinetics for hydrothermal and dilute acid pretreatments is important to choose the appropriate reactor configurations and operating conditions and design the reactor. However, pretreatment processes are complex in that they involve destruction and hydrolysis of a highly heterogeneous matrix of plant cell wall components including hemicelluloses that are solids in a liquid medium. Furthermore, although hemicellulose hydrolysis can be explained by a simple first-order kinetic model,[54] the kinetics have been observed to vary with reaction temperature.[59] For temperatures above 180 °C, hemicellulose decomposition into xylose has been described by a first-order reaction with kinetic rate parameter k_1 followed by xylose degradation via another first-order reaction with kinetic rate constant k_2:

$$\text{Hemicellulose} \xrightarrow{k_1} \text{Xylose} \xrightarrow{k_2} \text{Degradation products} \quad (3)$$

However at lower temperatures, hemicellulose hydrolysis is considered biphasic, with hemicellulose considered to be composed of two fractions, fast-hydrolyzing and slow-hydrolyzing, that release xylose via first-order reactions with rate constants k_f and k_s, respectively. Xylose once again degrades via a first-order reaction with rate constant k_2.[59,70] Fitting this model described in equation 4 results in about 20 to 35% of the xylan being categorized as slow hydrolyzing, while the remaining as fast hydrolyzing.[54]

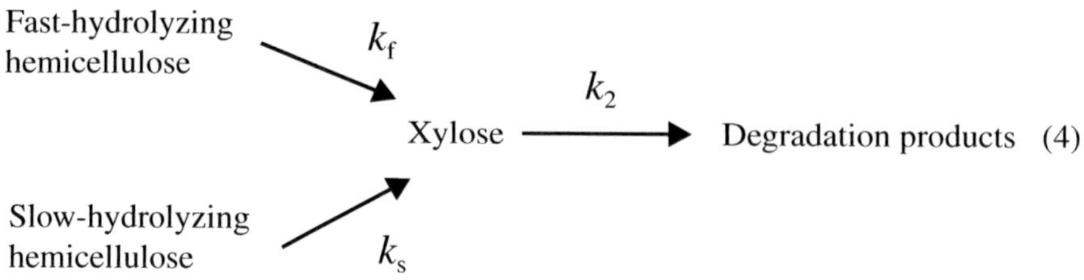

An improved version of this model introduced an additional hydrolysis step in which xylose oligomers are formed by a first-order reaction of the fast and slow reacting fractions of hemicellulose followed by a first-order reaction of the oligomers to xylose monomers with a rate constant k_0, leading to final reaction of xylose degradation by another first-order reaction.[54,59,71,72] This sequence thus becomes:

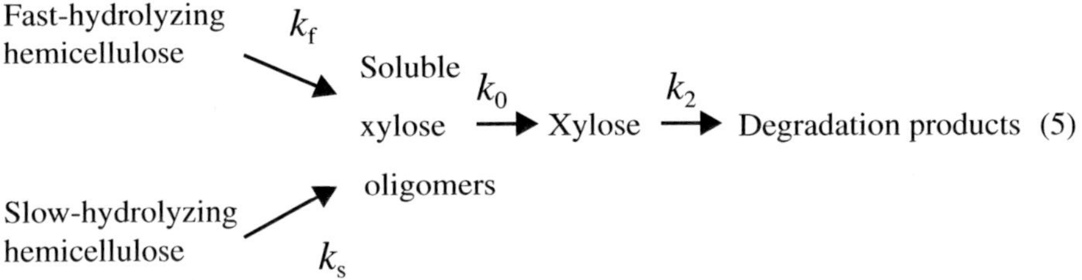

$$(5)$$

Direct degradation of xylose oligomers to unknown products has also been reported, with the rate primarily affected by pH.[73]

The kinetic rate parameters for this oligomer intermediate model can be calculated by applying Arrhenius relationships and acid–base catalysis. Since hydrothermal and acid pretreatments occur at low pH, the term representing hydroxyl ions can be ignored, leaving the following equation:[72]

$$k_i = (k_i^0 + k_i^H[10 - \text{pH}]) \exp\left(\frac{-E_i}{RT}\right) \tag{6}$$

Thus, only three parameters can be varied to optimize sugar release: temperature, pH, and time.

Overend and Chornet[13] introduced the severity parameter concept to provide a single reaction ordinate, denoted as the 'P' factor, that could combine time and temperature in a single parameter related to the changes in concentrations of the constitutive polymers in lignocellulosic biomass during autohydrolysis. The prototype of the reaction ordinate was developed earlier by Brasch and Free[74] and was well established in the pulp and paper industry as the 'H' factor. The reaction ordinate (or severity factor) has proven to be extremely useful in mapping hemicellulose sugar release from complex heterogeneous lignocellulosic materials and allowing easy determination of tradeoffs between time and temperature to get similar end results. This severity parameter, R_0, can be readily calculated using the following equation:

$$R_0 = t \exp\left(\frac{T - T_0}{14.75}\right) \tag{7}$$

where T is the reaction temperature (in °C), T_0 is the reference temperature (generally 100 °C), and t is the reaction time (in minutes). This equation can be shown to result from integration of the first-order reaction expression for hemicellulose hydrolysis while not separating the rate constant from the time variable and replacing the Arrhenius temperature dependence with the exponential term shown. The numerical value of 14.75 can be considered analogous to a conventional energy of activation by assuming that pretreatment is hydrolytic and xylan solubilization and lignin reduction follow first-order kinetics and obey the Arrhenius dependence on temperature. Alternatively, we can view this temperature dependence closely parallel to the heuristic observation that the reaction rate doubles for every $10\,^{\circ}\mathrm{C}$ increase in temperature. Typically, yield and other performance data are plotted against the logarithm of R_0 to compress the range of severity parameter values. If the temperature of the biomass material is subjected to a non-isothermal heat-up period, which can be particularly important if the heat up time is slow, the severity factor is more accurately calculated from:

$$R_0' = \int_0^t \left[\exp\left(\frac{T - 100}{14.75} \right) \right] \mathrm{d}t \tag{8}$$

The severity parameter assumes a constant pH during pretreatment, which can be reasonable for hydrothermal hydrolysis. However, because adding dilute acid will change the pH, the calculation of the severity parameter was extended to include the effect of acid addition by subtracting the pH from the $\log R_0$:[75,76]

$$\log CS = \log R_0 - \mathrm{pH} \tag{9}$$

Examples of the applicability of the severity parameter have been presented extensively[13] for such applications as tracking the recovery of pentosans, extractable lignin, and residual cellulose, and it has proven to be a valuable indicator of pretreatment process performance. The ability to relate (i) sugar yields in the liquid phase and (ii) cellulose digestibility in the solid residue following pretreatment to severity is particularly important in describing pretreatment processes. Linear or quadratic models based on severity parameter have been employed to explain the relationship between pretreatment severity and lignin reduction or xylan solubilization. Furthermore, severity parameter permits correlation of data obtained from different reactors operated at different conditions when the temperature–time history is known.

High yields of glucose and hemicellulose sugars from cellulose and hemicellulose in lignocellulosic biomass through the combined operations of pretreatment and enzymatic hydrolysis are essential for commercial processes based on biological conversion of these sugars to fuels or chemicals. However, as illustrated in Figure 4.5, the optimal pretreatment severity for xylose recovery is usually not the same as the optimal severity for the highest glucose yields during subsequent enzymatic hydrolysis of the pretreated cellulose.[27] A two-step steam pretreatment, with the first at low severity to hydrolyze hemicellulose and the second step at higher severity to enhance the digestibility of the solids from the first pretreatment step, can result in higher overall sugar yields than a one-step steam pretreatment process.[77] However, high capital and energy requirements along with an additional solid/liquid separation process

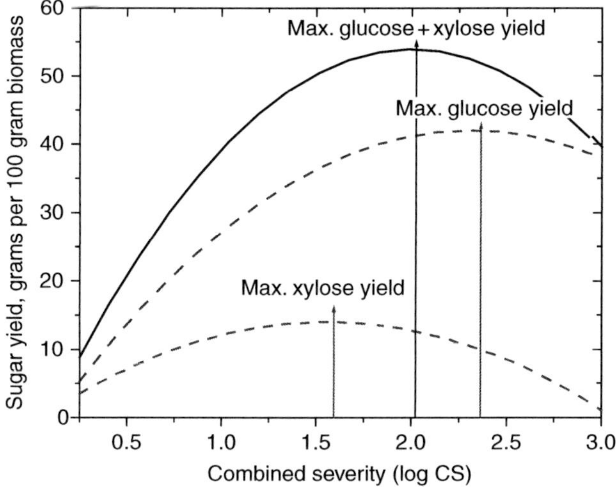

Figure 4.5 Typical sugar yield patterns expected from dilute acid pretreatment (lower curve) and enzymatic hydrolysis (middle curve) of hardwood as a function of the log of the combined severity parameter.

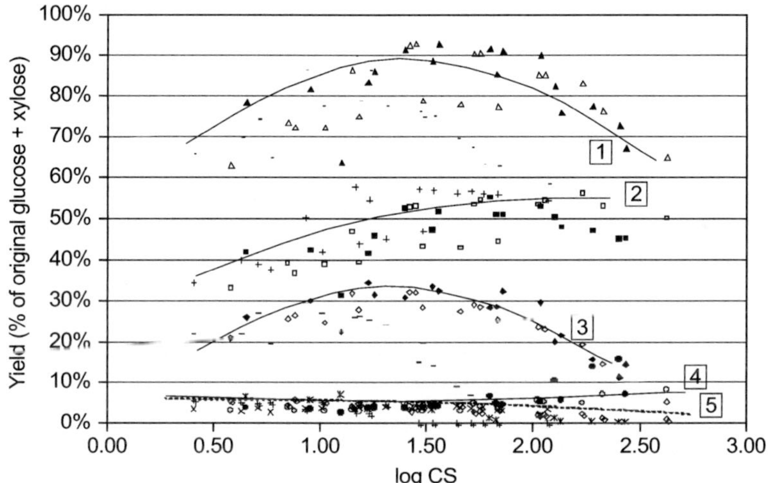

Figure 4.6 Component sugar yields vs. log combined severity parameter during dilute acid pretreatment (Stage 1) and enzymatic hydrolysis (Stage 2) of corn stover. Trend lines 1–5 are total Stage 1 and Stage 2 sugar, Stage 2 glucose, Stage 1 xylose, Stage 1 glucose, and Stage 2 xylose, respectively.[27]

between the two pretreatment steps limit the commercial viability of a two-step pretreatment.[78] Thus, compromise is generally required to obtain the highest possible yields from the combined operations based on one step pretreatment, as illustrated for Figure 4.6 for overall sugar release from dilute acid pretreatment and enzymatic hydrolysis of corn stover.[27] Although xylose release during pretreatment and glucose release during enzymatic hydrolysis are widely considered as the only yields of the sugars, an important realization that the release of some glucose during pretreatment and some xylose during enzymatic hydrolysis should also be factored in when defining optimal operating conditions came recently, as shown in Figure 4.6.[27]

4.5 Pretreatment Effects on the Digestibility of Post-pretreatment Solids

A vital goal of pretreatment is to open up the structure of lignocellulosic biomass and allow better access of cellulose to hydrolytic enzymes.[30,79–86] Cellulase adsorption on cellulose (accessibility) and their hydrolysis ability after they are adsorbed (effectiveness) are considered to be the main factors affecting enzymatic saccharification of cellulosic biomass.[87–90] However, both enzyme adsorption onto solids and their further effectiveness in hydrolysis can be strongly affected by substrate and enzymes characteristics, and physical parameters. The digestibility of post-pretreatment solids in turn is controlled by pretreatment efficacy and substrate features.[90] The major chemical and physical changes to lignocellulosic biomass by hydrothermal and dilute acid pretreatment are often attributed to the removal of hemicellulose, primarily xylan, and lignin modification and re-location. Other physiochemical changes associated with the major effects include disruption of biomass inter/intra linkages, increased surface area, enhanced pore volume due to xylan removal and lignin immigration, reduced cellulose degree of polymerization, and decreased biomass crystallinity.[30,56,79–84,86,91]

For some time, xylan removal was thought to improve cellulose accessibility by enzymes,[89,92] and new evidence supports the hypothesis that removing xylan could reduce strong inhibition of cellulase by xylooligomers,[93] leading to better digestibility of pretreated solids. Furthermore, a reduction in cellulose chain length accompanying xylan removal during dilute acid and hydrothermal pretreatment may also contribute to the improvement on cellulose digestion.[90] Additional evidences provided by FT-IR data indicated that dilute acid pretreatment of poplar increased the crystalline cellulose polymorphs ratio ($I\alpha/I\beta$) during pretreatment by $\sim$5.8%, partially contributing to enhanced digestibility.[94] Also, removal of acetyl content has been reported to enhance cellulose accessibility[92,95,96] and cellulase effectiveness for corn stover hydrolysis with purified CBHI.[90,93] In addition, acetyl removal may relieve the extent of inhibition of xylanase and debranching enzymes.[97] As suggested in a recent study by Selig and co-workers,[98] acetyl removal led to greater synergism between acetyl xylan esterase and cellobiohydrolase (CBHI), thus enhancing digestibility of post-pretreatment solids.

Two aspects implicate lignin as a major contributor to biomass recalcitrance and a barrier to enzymatic digestion: (i) its aromatic and hydrophobic nature, when exposed on biomass surface, impedes mass diffusion and enzyme accessibility and (ii) its ability to absorb cellulase enzymes unproductively, reduces enzyme effectiveness. Reduction in surface lignin during dilute acid and hydrothermal pretreatment, as measured by ESAC analysis, leads to a decrease in surface oxygen to carbon ratios, increase in cellulose exposure, and a reduction in surface hydrophobicity, all of which benefit digestibility.[91] In summary, the removal of either xylan or lignin and the associated physiochemical changes all serve one purpose: 'disruption of carbohydrate–lignin networking, which enhances enzymes adsorption, generally labeled as accessibility, and their effectiveness due to reduced inhibition by xylooligomers and unproductive binding with lignin.'[91]

4.6 Feedstock Considerations

Biorefineries will likely require more than one type of feedstock, and pretreatment processes do not affect all feedstocks in the same fashion even when treated at the same conditions because lignocellulosic biomass varies considerably in composition depending on feedstock type, breed, geographic location, harvest time, age, and environmental factors.[99,100] Thus, it is important to understand the effects of feedstock variability on pretreatment in addition to understanding the pretreatment process itself. In general, stalks (corn stover, baggasse, etc.) and straw (rice straw, wheat straw, etc.) require lower pretreatment severities to reach maximum sugar recoveries than woody materials, especially softwoods.[101] Furthermore, softwoods are more difficult to pretreat than hardwoods, and the available data suggests that catalysts such as sulfur dioxide are needed to achieve high digestibility.[57] Recent development of comparative data on sugar yields by the Biomass Refining Consortium for Applied Fundamentals and Innovation (CAFI) project demonstrated that both hydrothermal and dilute acid pretreatments realize higher sugar yields from corn stover than from low-lignin poplar, which in turn realized higher yields than from high-lignin poplar.[3,4,102] However, SO_2 catalyzed steam explosion was shown to be effective on all three of these feedstocks.[91,103]

Lignocellulosic biomass feedstocks contain significant amounts of minerals, usually characterized as ash, over a range of 0.5 to nearly 20 wt%, depending on species, harvesting operations, and location. The minerals in the biomass can potentially neutralize acid and reduce hydrogen ion activity and, thereby, reduce the pretreatment severity. Specific to sulfuric acid is an equilibrium shift, through which formation of bisulfates (HSO_4^-) and sulfates (SO_4^{2-}) reduces the hydrogen ion concentration.[104] Furthermore, bisulfate salt formation is favored in acidic solutions subjected to high temperatures, worsening the detrimental effects on pretreatment.[105] This neutralization plus shift effect can particularly affect pretreatment at low acid concentrations or high solids loadings, and predictions of the performance of hydrothermal and dilute acid pretreatments must take such effects into account to obtain accurate results.[104]

As a general rule, agricultural residues and herbaceous substrates, such as corn stover and switchgrass, respectively, have higher ash contents than poplar and other woody materials, and several studies demonstrate that higher ash contents result in greater neutralization of acid.[62] A dry ashing method developed by Springer and Harris[106] can be used to measure the neutralizing capacity of ash in lignocellulosic biomass by determining the amount of 0.005 M H_2SO_4 required to titrate biomass ash to a pH of 7. On this basis, the neutralizing abilities of corn stover, switchgrass, and poplar were determined to be 43.7, 25.8 and 16.7 mg H_2SO_4/g dry substrate, respectively.[62] An investigation by Maloney *et al.*[107] showed that poplar had a considerably higher neutralizing ability (16.7 mg H_2SO_4/g dry wood) than other hardwoods, i.e. paper birch (3.5 mg H_2SO_4/g dry wood) and red oak (8–9 mg H_2SO_4/g dry wood). In another study, the neutralizing capacity of the corn stover substrate was determined to be 17.3 mg H_2SO_4/g dry corn stover.[104] Neglecting the neutralization capacity of biomass can result in lower hemicellulose sugar yields,[62,108] making removal of minerals or neutralization products prior to hydrolysis very desirable, especially when processing high ash feedstocks at high solid loadings.

Addition of greater amounts of acid can compensate for the neutralization capacity of minerals in the feedstock,[67,109] and the amount of additional acid required to achieve optimal conditions can be calculated based on the neutralization capacity.[110,111] Lloyd and Wyman[112] suggested that less acid will be needed to achieve the same results if the minerals are removed from biomass prior to pretreatment by washing with dilute sulfuric acid. In addition to lowering the neutralization capacity of the solids, this washing approach reduces bisulfate formation whose buffering effect would require significantly higher quantities of acid to achieve satisfactory results.

4.7   Comparison of Hydrothermal and Dilute Acid Pretreatment Performance

Most of the xylan in lignocellulosic biomass can be released in both dilute acid and hydrothermal pretreatments, with monomers prevalent for dilute acid treatment and oligomers dominant for hydrothermal pretreatment. Table 4.1 provides a direct comparison between acid and hydrothermal pretreatments for corn stover with a composition of 22.6% xylan, 35.2% glucan, and 20.6% lignin.[3] Stage 1 refers to pretreatment and Stage 2 to enzymatic digestion of the solids produced by pretreatment. Yields are defined based on the maximum potential sugars released from the corn stover used, i.e. 64.4 g per 100 g of dry solids with a maximum xylose theoretical yield of about 37.7% and a maximum glucose theoretical yield of 62.3%. The overall xylose yield from dilute acid pretreatment is higher than from hydrothermal pretreatment, and both pretreatments hydrolyze only a small fraction of cellulose, with most of the glucose being released during enzymatic hydrolysis of cellulose. At an enzyme loading of 15 FPU/g glucan in the original corn stover, both pretreatments can produce total glucose yields close to the theoretical maximum. Thus, both pretreatments are effective in hydrolyzing hemicellulose and improving cellulose accessibility to enzymes. However, dilute acid realizes slightly higher glucose yields than

Table 4.1 Yields of xylose and glucose for dilute acid and hydrothermal pretreatment followed by enzymatic hydrolysis with a loading of 15 FPU/g glucan in the original corn stover.[3]

Pretreatment system		Dilute acid[27]		Hydrothermal with controlled pH[2]	
Pretreatment conditions		160 °C, 20 min, 0.5 wt% H_2SO_4		190 °C, 15 min	
		Total sugar yield (wt%)	Monomeric sugar yield (wt%)	Total sugar yield (wt%)	Monomeric sugar yield (wt%)
xylose yields	stage 1	32.1	31.2	21.8	0.9
	stage 2	3.2	3.2	9	9
	total xylose	35.3	34.4	30.8	9.9
glucose yields	stage 1	3.9	3.9	3.5	0.2
	stage 2	53.2	53.2	52.9	52.9
	total glucose	57.1	57.1	56.4	53.1
total sugars	stage 1	36.0	35.1	25.3	1.1
	stage 2	56.4	56.4	61.9	61.9
	combined total	92.4	91.5	87.2	63.0

hydrothermal pretreatment water and produces mostly monomeric sugars. Regarding total sugar yield from both pretreatments and enzymatic hydrolysis at high enzyme loadings, dilute acid realizes higher overall sugar yields of around 92% while hydrothermal obtains lower yields of about 87% and less. However, the enzyme doses used to obtain high yields in terms of mass of enzyme per volume of ethanol produced need to be reduced further to make biorefining economically feasible with either pretreatment.

4.8 Pretreatment Economics

Figure 4.7 provides an overview of the primary operations chosen for biological processing of lignocellulosic biomass to ethanol and also illustrates the interactions among individual processes.[64] Since pretreatment has such a pervasive effect on all other biorefinery operations, understanding pretreatment can accelerate selection of optimal process parameters, improve yields, and reduce costs of all operations significantly.[113,114] The various unit processes involved in pretreatment are given in Figure 4.8. In the case of dilute acid pretreatment, lignocellulosic biomass is milled and cleaned prior to soaking in about 1% sulfuric acid, and a solids concentration of 30% by weight is fed to a continuous plug flow pretreatment or prehydrolysis reactor.[115] Then, saturated steam is injected into the reaction vessel to rapidly raise the temperatures to between 170 °C and 200 °C while minimizing dilution of the sugars released during pretreatment due to the high heat of condensation of steam.

As shown in Figure 4.8, the solids and liquids are separated after pretreatment, and enough lime is added to the liquid to raise the pH to 10.0 in an

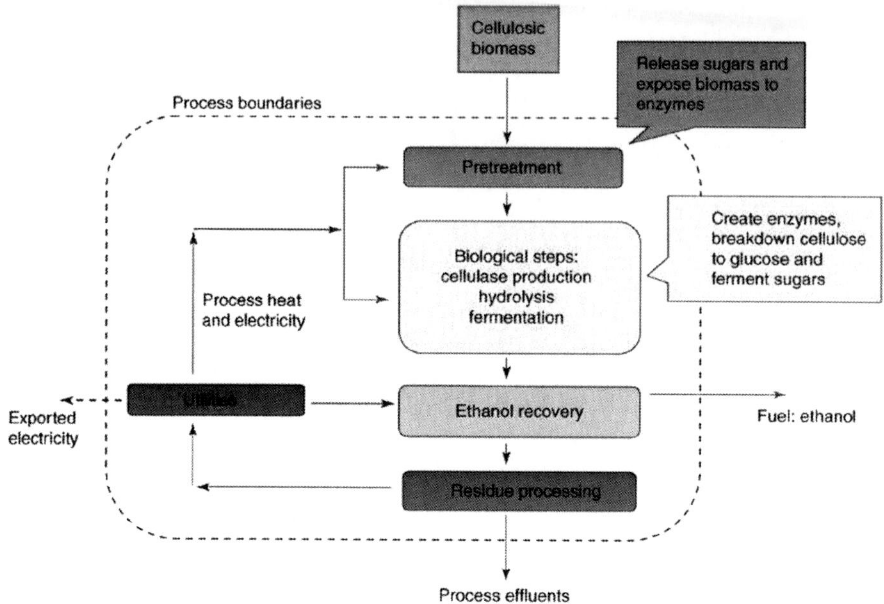

Figure 4.7 Process flow sheet for biological conversion of lignocellulosic biomass to ethanol.[114]

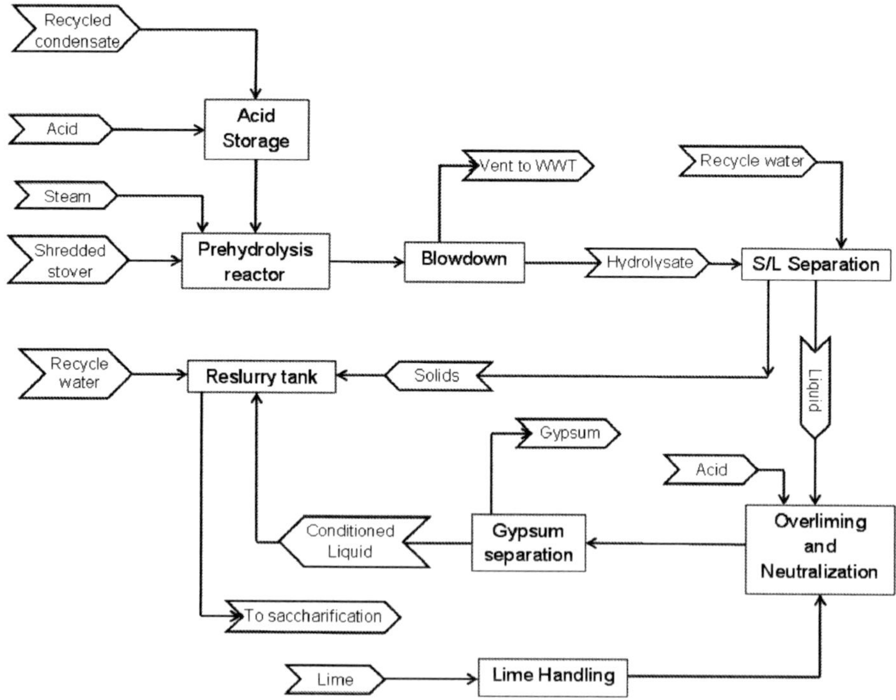

Figure 4.8 Overview of the operations directly associated with pretreatment of lignocellulosic biomass.[64]

overliming method that removes enough inhibitors to accommodate downstream biological operations. Gypsum formed by neutralization of acid and overliming is considered a waste and disposed in an appropriate manner.[116] The liquid is re-acidified with sulfuric acid to bring the pH to the desired levels such that high yields are obtained in the subsequent steps of enzymatic hydrolysis and fermentation.[115]

Yields will dominate the economics of producing commodity products from other than very low cost feedstocks and must therefore be as high as possible to achieve low unit product costs for large scale implementation of lignocellulosic conversion technologies. Once high yields are achieved, the costs of all unit operations must be as low as possible. Several studies have sought to estimate the economics of biological processing to ethanol and other products,[5,64,115–118] and many show pretreatment to be the most costly single unit process. For example, a recent evaluation of biorefinery economics estimated pretreatment costs to be the highest of all processing operation, accounting for about 14.6% of the cash costs, with only the cost of feedstock contributing more.[115]

4.9   Conclusions

Hydrothermal and dilute acid pretreatments can be effective in producing sugars from hemicellulose along with a solid residue enriched in cellulose that can be digested to glucose with high yields. Generally speaking, dilute acid pretreatment is often favored because it realizes higher yields than hydrothermal pretreatment and produces mostly monomeric sugars, but hydrothermal pretreatment can reduce the consequences of hydrolyzate conditioning to remove inhibitors, employ lower cost reaction vessels, and avoid the challenges of acid addition. The performance of hydrothermal systems correlates well with the severity parameter, while the modified severity parameter is an effective tool in analyzing dilute acid performance. Kinetic models have also been applied to describe sugar release profiles from dilute acid pretreatments, with the parameters fit to match the data, but the models are not robust in terms of a priori predictions of performance. Although cellulose digestion is enhanced significantly by pretreatment to the extent that virtually complete conversion of the cellulose is possible only in pretreated solids, the mechanism that governs glucose release by enzymes is not entirely clear due to the complexity of the substrate and enzyme action. In addition, the enzyme doses needed to realize high yields are excessive and not economically viable, and advances are sorely needed in the coupled operations of pretreatment and enzymatic hydrolysis to make high yields possible with much lower enzyme loadings. Recent studies have shown the importance of considering hemicellulose and cellulose sugar release during both pretreatment and enzymatic hydrolysis when identifying optimal pretreatment conditions to be applied and suggested not to focus solely on one step or one sugar. Feedstock features can have a significant effect on performance, with lignin amounts and mineral contents having potentially large effects. Economic studies clearly show that pretreatment is an expensive operation with pervasive impacts on the costs of other steps. Thus, although dilute acid and hydrothermal pretreatments appear to be frontrunners currently, much more must be done to understand and advance pretreatment technologies to realize low costs and high yields that are essential to production of commodity products.

Acknowledgements

Support by the U.S. Department of Energy Office of the Biomass Program (contract DE-FG36-04GO14017) made research possible to develop comparative pretreatment data through the CAFI project. We are also grateful to the Center for Environmental Research and Technology of the Bourns College of Engineering at the University of California, Riverside, for providing key equipment and facilities. The corresponding author is grateful to the Ford Motor Company for funding the Chair in Environmental Engineering at the Center for Environmental Research and Technology at the Bourns College of Engineering at UCR that makes possible many projects such as this one. We also thank Dr. Rajeev Kumar for his helpful input to this chapter.

References

1. M. E. Himmel, S.-Y. Ding, D. K. Johnson, W. S. Adney, M. R. Nimlos, J. W. Brady and T. D. Foust, *Science*, 2007, **315**, 804.
2. N. Mosier, C. E. Wyman, B. Dale, R. T. Elander, Y. Y. Lee, M. Holtzapple and M. Ladisch, *Bioresour. Technol.*, 2005, **96**, 673.
3. C. E. Wyman, B. E. Dale, R. T. Elander, M. Holtzapple, M. R. Ladisch and Y. Y. Lee, *Bioresour. Technol.*, 2005, **96**, 2026.
4. C. E. Wyman, B. E. Dale, R. T. Elander, M. Holtzapple, M. R. Ladisch, Y. Y. Lee, C. Mitchinson and J. N. Saddler, *Biotechnol. Progr.*, 2009, **25**, 333.
5. R. Wooley, M. Ruth, J. Sheehan, K. Ibsen, H. Majdeski and A. Galvez, *NREL/TP*, 1999, 580–26157, 1.
6. R. Boehm, *Ind. Eng. Chem.*, 1930, **22**, 493.
7. P. Foody, *Final Report to DOE*, 1980, Contract AC02-79ET23050, 1.
8. T. Schultz, C. Blermann and G. Mcginnis, *Ind. Eng. Chem. Prod. Res. Develop.*, 1983, **22**, 344.
9. O. Bobleter, R. Niesner and M. Rohr, *J. Appl. Polym. Sci.*, 1976, **20**, 2083.
10. G. Bonn, R. Concin and O. Bobleter, *Wood Sci. Technol.*, 1983, **17**, 195.
11. M. Wayman and J. Lora, *Tappi*, 1979, **62**, 113.
12. O. Bobleter, H. Binder, R. Concin and E. Burtscher, *Energy from Biomass*, ed. W. Palz, P. Chartier and D. O. Hall, Applied Science Publishers, London, 1981, p. 554.
13. R. Overend and E. Chornet, *Philos. Trans. R. Soc. London Ser. A*, 1987, **321**, 523.

14. J. Bouchard, T. S. Nguyen, E. Chornet and R. P. Overend, *Bioresour. Technol.*, 1991, **36**, 121.
15. W. S. L. Mok and M. J. Antal, *Ind. Eng. Chem. Res.*, 1992, **31**, 1157.
16. S. G. Allen, L. C. Kam, A. J. Zemann and M. J. Antal, *Ind. Eng. Chem. Res.*, 1996, **35**, 2709.
17. W. L. Faith, *Ind. Eng. Chem.*, 1945, **37**, 9.
18. E. E. Harris and E. Beglinger, *Ind. Eng. Chem.*, 1946, **38**, 890.
19. J. W. Dunning and E. C. Lathrop, *Ind. Eng. Chem.*, 1945, **37**, 24.
20. G. T. Tsao, M. R. Ladisch, M. Voloch and P. Bienkowski, *Process Biochem.*, 1982, **17**, 34.
21. S. M. Selke, M. C. Hawley, H. Hardt, D. T. A. Lamport, G. Smith and J. Smith, *Ind. Eng. Chem. Prod. Res. Dev.*, 1982, **21**, 11.
22. T. A. Hsu, M. R. Ladisch and G. T. Tsao, *Chem. Technol.*, 1980, **10**, 315.
23. H. E. Grethlein. *US Patent* 4,237,226, 1980.
24. D. Knappert, H. Grethlein and A. Converse, *Biotechnol. Bioeng.*, 1980, **22**, 1449.
25. B. Yang and C. E. Wyman, *Biotechnol. Bioeng.*, 2004, **86**, 88.
26. R. Datar, J. Huang, P. C. Maness, A. Mohagheghi, S. Czernik and E. Chornet, *Int. J. Hydrogen Energy*, 2007, **32**, 932.
27. T. A. Lloyd and C. E. Wyman, *Bioresour. Technol.*, 2005, **96**, 1967.
28. L. Ramos, C. Breuil, D. Kushner and J. Saddler, *Holzforschung*, 1992, **46**, 149.
29. C. Tengborg, K. Stenberg, M. Galbe, G. Zacchi, S. Larsson, E. Palmqvist and B. Hahn-Hagerdal, *Appl. Biochem. Biotechnol.*, 1998, **70**, 3.
30. J. N. Saddler, H. H. Brownell, L. P. Clermont and N. Levitin, *Biotechnol. Bioeng.*, 1982, **24**, 1389.
31. R. H. Marchessault and J. St-Pierre, *Future Sources of Organic Raw Materials – CHEMRAWN I*, ed. L. E. St. Pierre and G. R. Brown, Pergamon Press, New York, 1980, p. 613.
32. D. A. I. Goring, *Pulp and Paper Magazine of Canada*, 1963, **64**, 517.
33. B. S. Donohoe, S. R. Decker, M. P. Tucker, M. E. Himmel and T. B. Vinzant, *Biotechnol. Bioeng.*, 2008, **101**, 913.
34. C. Liu and C. E. Wyman, *Ind. Eng. Chem. Res.*, 2003, **42**, 5409.
35. M. G. S. Chua and M. Wayman, *Can. J. Chem.*, 1979, **57**, 1141.
36. M. Bardet, M. F. Foray and D. Robert, *Macromol. Chem. Phys.*, 1985, **186**, 1495.
37. H. L. Chum, D. K. Johnson, M. P. Tucker and M. E. Himmel, *Holzforschung*, 1987, **41**, 97.
38. R. H. Marchessault, S. L. Malhatra, A. Y. Jones and A. Perovic, in *Wood and Agricultural Residues*, Academic Press, New York, London, 1983, p. 401.
39. E. Back and N. Salmen, *Tappi*, 1982, **65**, 107.
40. E. Palmqvist, B. Hahn-Haegerdal, M. Galbe and G. Zacchi, *Enzyme Microb. Technol.*, 1996, **19**, 470.
41. E. Palmqvist E and B. Hahn-Hagerdal, *Bioresour. Technol.*, 2000, **74**, 17.

42. R. H. Leonard and G. J. Hajny, *Ind. Eng. Chem.*, 1945, **37**, 390.
43. J. Zaldivar and L. O. Ingram, *Biotechnol. Bioeng.*, 1999, **66**, 203.
44. H. G. Lawford, J. D. Rousseau, A. Mohagheghi and J. D. McMillan, *Appl. Biochem. Biotechnol.*, 1999, **77**, 191.
45. E. Casey, M. Sedlak, N. W. Ho and N. S. Mosier, *FEMS Yeast Research*, 2010, **10**, 385.
46. N. Banerjee, R. Bhatnagar and L. Viswanathan, *Appl. Microbiol. Biotechnol.*, 1981, **11**, 226.
47. J. J. Heipieper, F. J. Weber, J. Sikkema, H. Keweloh and J. A. M. de Bon, *Trends Biotechnol.*, 1994, **12**, 409.
48. J. Zaldivar, A. Martinez and L. O. Ingram, *Biotechnol. Bioeng.*, 1999, **65**, 24.
49. A. Martinez, M. E. Rodriguez, S. W. York, J. F. Preston and L. O. Ingram, *Biotechnol. Bioeng.*, 2000, **69**, 526.
50. T. D. Ranatunga, J. Jervis, R. F. Helm, J. D. McMillan and R. J. Wooley, *Enz. Microb. Technol.*, 2000, **27**, 240.
51. P. Pienkos and M. Zhang, *Cellulose*, 2009, **16**, 743.
52. S. Amartey and T. Jeffries, *World J. Microb. Biotechnol.*, 1996, **12**, 281.
53. R. Wooley, M. Ruth, D. Glassner and J. Sheehan, *Biotechnol. Prog.*, 1999, **15**, 794.
54. Y. Y. Lee, P. Iyer and R. W. Torget, *Adv. Biochem. Eng. Biotechnol.*, 1999, **65**, 93.
55. M. Studer, J. DeMartini, S. Brethauer, H. McKenzie and C. E. Wyman, *Biotechnol. Bioeng.*, 2010, **105**, 231.
56. M. P. Tucker, R. K. Mitri, F. P. Eddy, Q. A. Nguyen, L. M. Gedvilas and J. D. Webb, *Appl. Biochem. Biotechnol.*, 2000, **84**, 39.
57. Q. A. Nguyen, M. P. Tucker, B. L. Boynton, F. A. Keller and D. J. Schell, *Appl. Biochem. Biotechnol.*, 1998, **70**, 77.
58. S. L. Stuhler and C. E. Wyman, *Appl. Biochem. Biotechnol.*, 2003, **105**, 101.
59. S. E. Jacobsen and C. E. Wyman, *Appl. Biochem. Biotechnol.*, 2000, **84**, 81.
60. J. S. Kim, Y. Y. Lee and R. W. Torget, *Appl. Biochem. Biotechnol.*, 2001, **91**, 331.
61. N. S. Mosier, C. M. Ladisch and M. R. Ladisch, *Biotechnol. Bioeng.*, 2002, **79**, 610.
62. A. Esteghlalian, A. Hashimoto, J. Fenske and M. Penner, *Bioresour. Technol.*, 1997, **59**, 129.
63. B. Yang and C. E. Wyman, *Biotechnol. Bioeng.*, 2006, **94**, 611.
64. A. Aden, M. Ruth, K. Ibsen, J. Jechura, K. Neeves, J. Sheehan, B. Wallace, L. Montague, A. Slayton and J. Lukas, *NREL/TP*, 2002, 510–32438, 1.
65. D. R. Thompson and H. E. Grethlein, *Ind. Eng. Chem.*, 1979, **18**, 166.
66. K. J. Zeitsch, *The Chemistry and Technology of Furfural and its Many By-products*, In: Sugar Series 13, Elsevier Science, Amsterdam, The Netherlands, 2000.
67. D. R. Cahela, Y. Y. Lee and R. P. Chambers, *Biotechnol. Bioeng.*, 1983, **25**, 3.

68. J. D. Wright, P. W. Bergeron and P. J. Werdene, *Ind. Eng. Chem. Res.*, 1987, **6**, 699.
69. R. W. Torget, T. K. Hayward and R. T. Elander. *19th Symposium on Biotechnology for Fuels and Chemicals*, Colorado Springs, CO, 1997.
70. T. Kobayashi and Y. Sakai, *Bull. Agric. Chem. Soc. Jpn.*, 1956, **20**, 1.
71. G. Garrote, H. DomõÂnguez and J. C. Parajo, *HolzalsRoh-und Werkstoff*, 1999, **57**, 191.
72. D. J. Schell, J. Farmer, M. Newman and J. D. McMillan, *Appl. Biochem. Biotechnol.*, 2003, **105**, 69.
73. R. Kumar and C. E. Wyman, *Carbohydr. Res.*, 2008, **343**, 290.
74. D. J. Brasch and K. W. Free, *Tappi*, 1965, **48**, 245.
75. N. Abatzoglou, E. Chornet, K. Belkacemi and R. Overend, *Chem. Eng. Sci.*, 1992, **47**, 1109.
76. H. L. Chum, D. K. Johnson, S. K. Black and R. P. Overend, *Appl. Biochem. Biotechnol.*, 1990, **24**, 1.
77. J. Soderstrom, L. Pilcher, M. Galbe and G. Zacchi, *Appl. Biochem. Biotechnol.*, 2002, **98**, 5.
78. A. Wingren, J. Soderstrom, M. Galbe and G. Zacchi, *Biotechnol. Prog.*, 2004, **20**, 1421.
79. T. Clark, K. Mackie, P. Dare and A. Mcdonald, *J. Wood Chem. Technol.*, 1989, **9**, 135.
80. G. Excoffier, B. Toussaint and M. Vignon, *Biotechnol. Bioeng.*, 1991, **38**, 1308.
81. H. E. Grethlein, *Biotechnol. Adv.*, 1984, **2**, 43.
82. W. R. Grous, A. O. Converse and H. E. Grethlein, *Enzyme Microb. Technol.*, 1986, **8**, 274.
83. G. Michalowicz, B. Toussaint and M. R. Vignon, *Holzforschung*, 1991, **45**, 175.
84. M. J. Selig, S. Viamajala, S. R. Decker, M. P. Tucker, M. E. Himmel and T. B. Vinzant, *Biotechnol. Prog.*, 2007, **23**, 1333.
85. M. P. Tucker, J. D. Farmer, F. A. Keller, D. J. Schell and Q. A. Nguyen, *Appl. Biochem. Biotechnol.*, 1998, **70**, 25.
86. K. K. Y. Wong, K. F. Deverell, K. L. Mackie, T. A. Clark and L. A. Donaldson, *Biotechnol. Bioeng.*, 1988, **31**, 447.
87. H. C. Chen and H. E. Grethlein, *Biotechnol. Lett.*, 1988, **10**, 913.
88. A. R. Esteghlalian, M. Bilodeau, S. D. Mansfield and J. N. Saddler, *Biotechnol. Prog.*, 2001, **17**, 1049.
89. T. Jeoh, C. I. Ishizawa, M. F. Davis, M. E. Himmel, W. S. Adney and D. K. Johnson, *Biotechnol. Bioeng.*, 2007, **98**, 112.
90. R. Kumar and C. E. Wyman, in *Bioalcohol Production: Biochemical Conversion of Lignocellulosic Biomass*, ed. K. Waldon, Woodhead Publishing Limited, Oxford, 2010, p. 73.
91. R. Kumar and C. E. Wyman, *Biotechnol. Prog.*, 2009, **25**, 302.
92. K. Grohmann, D. J. Mitchell, M. E. Himmel, B. E. Dale and H. A. Schroeder, *Appl. Biochem. Biotechnol.*, 1989, **20**, 45.
93. R. Kumar and C. E. Wyman, *Biotechnol. Bioeng.*, 2009, **102**, 457.

94. R. Kumar, G. Mago, V. Balan and C. E. Wyman, *Bioresour. Technol.*, 2009, **100**, 3948.

95. V. S. Chang and M. T. Holtzapple, *Appl. Biochem. Biotechnol.*, 2000, **84**, 5.

96. T. M. Wood and S. I. McCrae, *Phytochemistry*, 1986, **25**, 1053.

97. D. J. Mitchell, K. Grohmann, M. E. Himmel, B. E. Dale and H. A. Schroeder, *J. Wood Chem. Technol.*, 1990, **10**, 111.

98. M. J. Selig, E. P. Knoshaug, W. S. Adney, M. E. Himmel and S. R. Decker, *Bioresour. Technol.*, 2008, **99**, 4997.

99. A. Wiselogel, S. Tyson and D. Johnson D, *Handbook on Bioethanol: Production and Utilization*, ed. C. E. Wyman, Taylor & Francis, Washington DC, 1996, p. 105.

100. K. H. Han, J. H. Ko and S. H. Yang, *Biofuels Bioproducts Biorefining*, 2007, **1**, 135.

101. M. Galbe and G. Zacchi, *Adv. Biochem. Eng./Biotechnol.*, 2007, **108**, 41.

102. T. Richard, R. T. Elander, B. E. Dale, M. Holtzapple, M. R. Ladisch, Y. Y. Lee, C. Mitchinson, J. N. Saddler and C. E. Wyman, *Cellulose*, 2009, **16**, 649.

103. R. Bura, S. Mansfield, J. Saddler and R. Bothast, *Appl. Biochem. Biotechnol.*, 2002, **98**, 59.

104. T. A. Lloyd and C. E. Wyman, *Appl. Biochem. Biotechnol.*, 2004, **113**, 1013.

105. J. Readnour and J. Cobble, *Inorg. Chem.*, 1969, **8**, 2174.

106. E. Springer and J. Harris, *Ind. Eng. Chem. Prod. Res. Dev.*, 1985, **24**, 485.

107. M. Maloney, T. Chapman and A. Baker, *Biotechnol. Bioeng.*, 1985, **27**, 355.

108. I. K. Kwarteng, 1983, Ph.D. thesis, Thayer School of Engineering, Dartmouth College, Hanover, NH 03755, USA.

109. A. H. Conner, B. F. Wood, C. G. Hill and J. F. Harris, in *Cellulose: Structure, Modification, and Hydrolysis*, ed. R. A. Young and R. M. Rowe, Wiley, New York, 1986, p. 281.

110. K. D. Baugh, J. A. Levy and P. L. McCarty, *Biotechnol. Bioeng.*, 1988, **31**, 62.

111. I. A. Malester, M. Green and G. Shelef, *Ind. Eng. Chem. Res.*, 1992, **31**, 1998.

112. C. E. Wyman and T. A. Lloyd, *US Patent* 7503981, 2006.

113. C. E. Wyman, *Bioresour. Technol.*, 1994, **50**, 3.

114. C. E. Wyman, *Trends Biotechnol.*, 2007, **25**, 153.

115. A. Aden and T. Foust, *Cellulose*, 2009, **16**, 535.

116. C. N. Hamelinck, G. van Hooijdonk and A. P. C. Faaij, *Biomass and Bioenergy*, 2005, **28**, 384.

117. F. K. Kazi, J. A. Fortman, R. P. Anex, D. D. Hsu, A. Aden, A. Dutta and G. Kothandaraman, *Fuel*, 2010, **89**, S20.

118. L. Lynd, J. Cushman, R. Nichols and C. E. Wyman, *Science*, 1991, **251**, 1318.

CHAPTER 5

A Short Review on Ammonia-based Lignocellulosic Biomass Pretreatment

VENKATESH BALAN,[1,2] BRYAN BALS,[1,2]
LEONARDO DA COSTA SOUSA,[1] REBECCA
GARLOCK[1,2] AND BRUCE E. DALE[1,2]

[1] Biomass Conversion Research Lab (BCRL), Department of Chemical
Engineering and Materials Science, Michigan State University, 3900 Collins
Road, MBI International Building, Lansing, MI 48910, USA; [2] DOE Great
Lakes Bioenergy Research Center, Michigan State University

5.1 Introduction

Pretreatment is an important process operation within a biorefinery and is
performed to convert lignocellulosic biomass into fuels and chemicals while
affecting all other downstream operations including enzymatic hydrolysis and
fermentation (Figure 5.1). Different varieties of pretreatment technology are
available (incorporating acid, neutral, and alkaline pH conditions), with physically, chemically, and biologically based processes. Chemically based pretreatments are considered to be the most promising for future biorefineries when
compared to other available pretreatment processes. However, it is important to
evaluate technical, economical, and environmental issues before adapting these
technologies for the future biorefinery. A number of review articles have been
published that provide details about the available pretreatment methods and

RSC Energy and Environment Series No. 4
Chemical and Biochemical Catalysis for Next Generation Biofuels
Edited by Blake Simmons
© Royal Society of Chemistry 2011
Published by the Royal Society of Chemistry, www.rsc.org

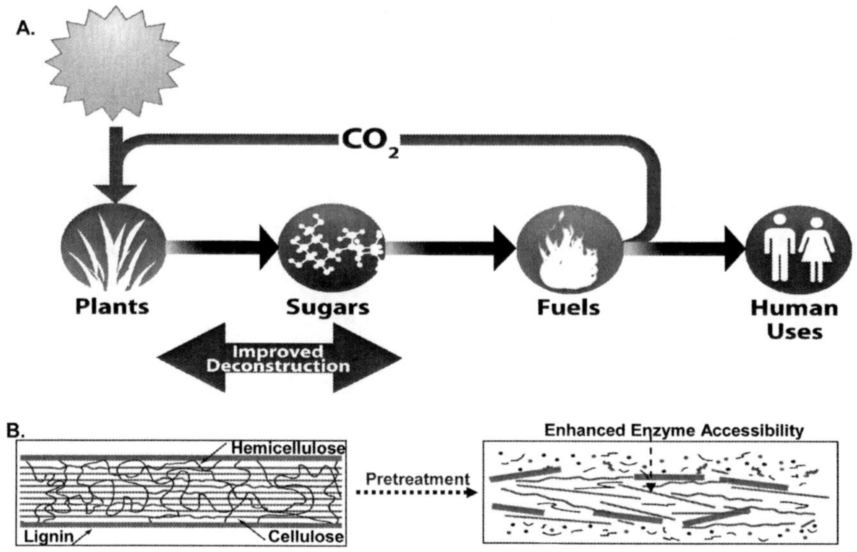

Figure 5.1   Picture showing various steps involved in the biofuel production process. Here (A) shows sustainable production of biofuels for human use from biomass and (B) lignocellulosic biomass showing the complex matrix of cellulose, hemicellulose, and lignin and how pretreatment helps to disrupt the complex structure for enhanced enzyme accessibility.

their advantages and disadvantages.[1–5] Economic analyses for some of the pretreatment processes and how they contribute to the minimum ethanol selling price have also been reported.[6,7]

5.2   Alkaline Pretreatment Processes

In many reported alkaline pretreatment processes, calcium oxide (lime), ammonia, or sodium hydroxide are used. Most of these chemicals specifically target hemicellulose acetyl groups and lignin–carbohydrate ester linkages.[8,9] These reactions help solubilize and extract lignin from the biomass, reducing non-specific binding during enzymatic hydrolysis. However, cell wall chemical and ultra-structural modifications still need to be understood for most alkaline pretreatments in order to develop proper enzyme mixtures that can effectively hydrolyze both cellulose and hemicellulose.

During alkaline pretreatment the initial reactions are solvation and saponification which cause the biomass to swell and make it more accessible to enzymes and biomass-degrading bacteria. Pretreatment using a stronger alkali, such as NaOH, at higher concentrations can result in dissolution and alkaline hydrolysis ('peeling' of end groups), eventually resulting in the degradation and decomposition of dissolved polysaccharides.[10,11] The monomers of hemicellulose may be easily degradable to other (volatile) compounds, which can lead to losses of digestible substrate for the ethanol process. An important aspect of alkali pretreatment is that the biomass itself consumes some of the alkali. For example, in the case of NaOH pretreatment (under specific conditions) 3 g of NaOH is consumed per 100 g total solids.[12] Alkali extraction can also cause solubilization, redistribution, and condensation of lignin as well as modifications to the crystalline state of the cellulose.[13,14]

Another important aspect of alkaline pretreatment is the change of the native cellulose structure to four different allomorphs.[15,16] The allomorphs of cellulose have been studied in detail using X-ray diffraction patterns,[17] NMR,[18] and vibrational spectroscopy.[19] Different allomorphs of cellulose can be prepared using a well defined protocol as summarized in Figure 5.2. All of the cellulose allomorphs are composed of layered sheets which contain intra-chain and inter-chain hydrogen bonding and, in some cases, these sheets may interact via hydrogen bonding.[17] Three of the cellulose allomorphs (I and its inter-convertible relatives III_I and IV_I) lack inter-sheet hydrogen bonds and are thought to display parallel sheet packing, while the other two (II and III_{II}) purportedly contain inter-sheet hydrogen bonds and display anti-parallel sheet packing.

The relative rates of digestion of the different cellulose allomorphs were digested using ruminal bacteria and were found to break down in the following order: amorphous > III_I > IV_I > III_{II} > I > II.[20] In one other study, cellobiohydrolase I (Cel7A) reactivity on different allomorphs of cellulose was

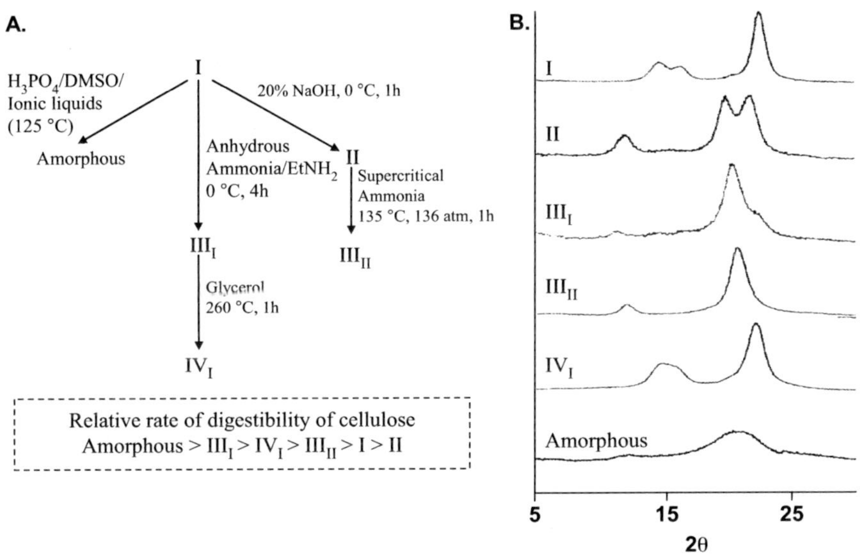

Figure 5.2 Different allomorphs of cellulose and how they can be distinguished based on their X-ray profile. Here (A) a scheme for converting cellulose I to other allomorphs using different chemicals and (B) powder X-ray diffraction spectra of different cellulose allomorphs. Figures adopted from Weimer *et al.*[20] with permission.

reported.[21] In this case, cellobiose production from cellulose III$_I$ was five times higher than that produced from cellulose I. However, the amount of enzyme adsorbed on cellulose III$_I$ was two times lower than that adsorbed on cellulose I, while the specific activity of the adsorbed enzyme for cellulose III$_I$ was three times higher than that for cellulose I. The possible reason for the higher rate of hydrolysis of cellulose III$_I$ by Cel7A is the higher specific activity of the enzyme for this crystalline polymorph and not due to the larger surface area of the substrate.[21] Cellulose III$_I$ can also be converted back to cellulose I either by reaction with glycerol or water at high temperatures. Based on the above-mentioned studies it is evident that converting cellulose I to amorphous cellulose or cellulose III$_I$ during pretreatment will result in highly digestible biomass.

5.2.1 Different Types of Alkali-based Pretreatment Processes

The concentration of catalyst used, catalyst recovery, and water usage for the most important alkaline-based pretreatment processes are summarized in Table 5.1. Several of these processes are discussed in detail in numerous review articles which have been published in recent years. These include lime pretreatment,[22] aqueous ammonia-based pretreatment processes including ammonia recycle percolation (ARP) and soaking in aqueous ammonia (SAA),[23] AFEX,[24] super-critical ammonia pretreatments,[25,26] ammonia–hydrogen peroxide pretreatments,[27] and alkaline wet oxidation, a form of oxidative pretreatment.[28,29] Overall, these pretreatments increase the biomass surface area by solubilization of hemicellulose and/or lignin and/or altering the lignin. Of the different alkalis used in the pretreatment process, ammonia is the only catalyst which is volatile and can be recovered and re-used, which has some technical advantages.

5.2.2 Ammonia and its Properties

Ammonia is a colorless gas with a distinct and pungent odor. It is synthesized using the Haber-Bosch process, which involves combining one volume of nitrogen with three volumes of hydrogen at temperatures as high as $1000\,°C$.[30] Ammonia can be liquefied due to the strong hydrogen bonding between molecules; the ammonia liquid boils at $-33.3\,°C$ and solidifies at $-77.7\,°C$ to white crystals. Ammonia is commercially used as anhydrous ammonia. In 2006 the world production was estimated to be 145.5 million tons. About 80% of the ammonia produced world-wide is used to make fertilizers (combining ammonia with the appropriate acids to form ammonium salts) for agricultural purposes.

5.2.3 History of Using Ammonia as a Pretreatment Chemical

The first reports on cellulose fiber modification using ammonia appeared in the 1930s.[31] Soon after, a number of other reports on the properties of pretreated biomass using different forms of ammonia were reported. Fiber separation for

Table 5.1 Different conditions adopted during alkaline-based biomass pretreatment processes.

Pretreatments	Catalyst	Recovery	Pressure (psig)	Temperature (°C)	Residence time	Reference
Aqueous ammonia						
ammonia recycled perculation	ammonia/water	up to 95%	250–350	160–180	10–20 min	23
soaking aqueous ammonia	ammonia/water	up to 95%	0	60	4–24 hours	23, 37
Concentrated ammonia						
super critical ammonia	ammonia	up to 98%	2000–4000	160–200	10 min–3 hours	25, 26
AFEX	ammonia/water	up to 98%	200–600	70–140	5–30 min	24
Sodium hydroxide						
concentrated	NaOH/water	can be	0	90	30 min	10, 71
dilute	NaOH/water	no	0	90	24 hours	10, 71
alkaline peroxide	NaOH/ammonia with H_2O_2/peracetic acid	no	0–300	60–120	6–24 hours	27
alkaline wet oxidation	water/oxygen and air or H_2O_2/Na_2CO_3	no	45–175	170–220	15 min	28, 29
Lime						
simple pretreatment	$Ca(OH)_2$/boiling water	possible	0	100	1 hour	22
short-term pretreatment	water/$Ca(OH)_2$/O_2	possible	200	100–160	6 hours	22
long-term pretreatment	water/$Ca(OH)_2$/air or O_2 (optional)	possible	0	55–65	8 weeks	22

pulping purposes by means of explosive decompression of wood chips with ammonia at elevated temperatures and pressures was also reported in the early 1970s.[32] For several decades, animal farmers have treated their animal feed using ammonia or urea, increasing the nitrogen content and digestibility of the feed and subsequently increasing livestock milk production. For this method, a stack of hay or grass on the field is covered with a polyethylene sheet to create an air- and ammonia-tight seal.[33] Anhydrous ammonia is then injected into the stack and allowed to react with the roughage for one to four weeks.[34] Though the digestibility of feed increases by up to 36%, this procedure is a labor intensive and uncontrolled way of producing highly digestible animal feed. The pretreatment can take up to several months and will vary depending on weather conditions. It also creates environmental issues by releasing excess ammonia into the environment. In the early 1980s, a novel lignocellulosic pretreatment method using ammonia was developed called the freeze-explosion method.[35] The process was later renamed ammonia fiber explosion and eventually patented in late 1980s.[36] In the middle of 2005 the process was again renamed ammonia fiber expansion (AFEX). With the help of this process, the biomass can be pretreated reliably in shorter periods of time at moderately elevated temperatures. The process is economically viable, with little impact on the environment since 97–98% of ammonia can be recovered and reused. In mid 1980s super critical ammonia pretreatment was developed.[25,26] Because of economical considerations that process had limited application on lignocellulosic biomass pretreatment. Recently, DuPont has developed a process of pretreating corn cobs using dilute ammonia.[37] Since the ammonia concentration is relatively low no recovery step is involved.

5.3 Details of the AFEX Process

AFEX is a novel alkaline pretreatment process that has been successful in improving the lignocellulosic biomass degradability by hydrolytic enzymes. During the AFEX process (Figure 5.3), the biomass with the desired moisture content is sealed in a pre-heated reactor at optimal reaction temperature. Concentrated anhydrous ammonia is added to the system. The ammonia dissociates in water to form ammonium and hydroxide ions, by the following equation:

$$NH_3 + H_2O \rightarrow NH_4^+ + OH^-$$

During this step, heat is released (ΔH of formation of NH_4OH is –87.59 kcal/mol at 25 °C) and used to rapidly increase the temperature of the biomass in the reactor.

When the reaction temperature is reached, the system is maintained with constant temperature and pressure for the desired residence time. After the residence time is reached, the pressure of the reactor is released, decreasing the temperature of the system to approximately 20–30 °C. Several types of biomass have been studied (details given below) and the optimal reaction parameters depend on their characteristics.

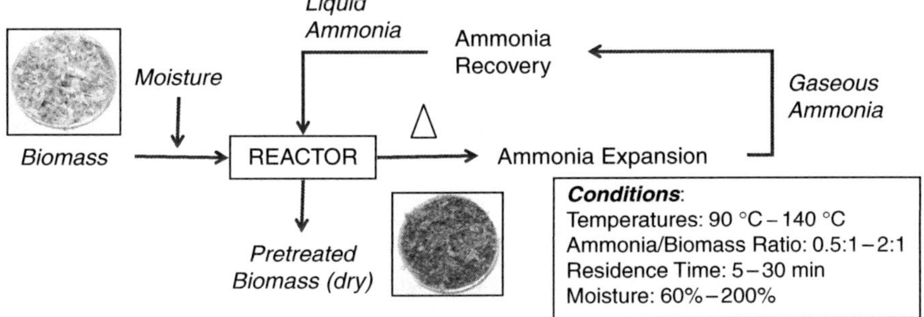

Figure 5.3   Picture demonstrating an AFEX pretreatment design where ammonia can be recovered and reused.

5.3.1   Pretreatment Variables

The important variables during the AFEX process are: reaction temperature, residence time, biomass moisture, and ammonia to dry biomass loading. In most of the pretreatment studies to optimize AFEX conditions for a given biomass, temperatures ranged from 70 °C to 180 °C, the ammonia to biomass loading varied from 0.6:1 to 2:1, moisture to biomass loading between 0.6:1 and 23:1, and the residence time varied from 15 to 30 minutes. AFEX has been shown to have tremendous potential to pretreat grasses such as corn stover and switchgrass, improving the degradability of the biomass 4–5 fold. These improvements on biological susceptibility are a result of physicochemical alterations of the lignocellulosic biomass ultra-structure during the AFEX process. These alterations are related to chemical interactions that occur in alkali conditions between ammonia and the biomass. The details about the AFEX reactor and detailed operations are described elsewhere.[24]

5.3.2   Fundamental Understanding of the Alkaline Pretreatment Process

Understanding alkaline degradation of polysaccharides, like cellulose and hemicellulose, is of extreme importance to the pulping industry and recently to the emerging lignocellulosic ethanol industry, since the degradation of these sugar polymers decreases the overall yield of both industrial processes. Initial reactions consist of the solvation of hydroxyl groups by hydroxyl ions. This phenomenon causes the biomass to swell. At higher temperatures, there are a large number of possible chemical reactions that can occur. The most important ones are considered here:[10] (i) dissolution of un-degraded polysaccharides; (ii) peeling of end groups with formation of alkali stable end groups; (iii) alkaline hydrolysis of glucosidic bonds and acetyl groups; (iv) degradation and decomposition of dissolved polysaccharides and peeled monosaccharides.

Loss of polysaccharides or a decrease in the degree of polymerization during alkali treatments are largely due to peeling and hydrolytic reactions. These are the most important reactions to track in order to preserve the integrity of the polysaccharides. At temperatures around 100 °C the degradation of polysaccharide chains begins in the existing reducing end groups, commonly called the primary peeling reaction. At temperatures around 150 °C alkaline hydrolysis of the chains begins and the newly formed reducing ends will also be subject to peeling reactions (secondary peeling). These secondary reactions are also called endwise peeling reactions.[9]

The peeling reaction of polysaccharides results in the elimination of the reducing end groups of polysaccharides, forming various carboxylic acid compounds. Sjöström[9] proposed a mechanism for endwise peeling where the initial step consists of the isomerization of the reducing end group to a ketose, which is in equilibrium with the corresponding 2,3-enediol. The C4-substituent is alkali labile in this conformation which promotes its cleavage, leading to the formation of a new reducing end in the polysaccharide chain. The resulting monomeric sugar can be tautomerized to a dicarbonyl compound that can then be rearranged to yield isosaccharinic acids. Other degradation products that can result from these reactions may be lactic acid, 2-hydroxybutanoic acid and 2, 5-dihydroxypentanoic acid.

Hemicelluloses are in general much more vulnerable to these types of chemical reactions in alkali media than celluloses. However, one can also see differences among the different kinds of hemicellulose components. Xylans are more stable and less vulnerable than glucomannans and arabinans. The easy cleavage of arabinose side groups in softwood xylans has a stabilizing effect against alkaline peeling, since an alkali-stable metasaccharinic acid end group is formed after the loss of the arabinose side group.[10]

The reaction that occurs during endwise peeling terminates when competing reactions begin to take place. These reactions are called stopping reactions and are very important to avoid the disruption of the polysaccharide fibers and their degradation. These reactions are initiated by a β-hydroxy elimination at the C2 position, producing a tautomeric intermediate that is converted to an alkali stable metasaccharinic acid group or C2-methylglyceric acid.[9]

Other small molecular weight acids are also formed in alkali conditions, like acetic acid and formic acid. Formic acid is a product of peeling reactions of polysaccharides, while acetic acid is formed by cleavage of the acetyl side chain groups present in grasses and hardwoods xylans. Glucuronic acid side groups of xylan are also hydrolyzed during intensive alkaline pretreatment.[9,10]

In any discussion about lignin degradation, the most important aspect is the varied behavior and stability of the possible bonds that exist in the lignin polymer. The most labile bonds in alkaline conditions are the aryl-ether bonds, which are mostly cleaved during kraft pulping treatment. Also, aryl-alkyl or alkyl-alkyl bonds are typically destroyed in alkaline medium, but to a lesser extent than aryl-ether bonds. Diaryl ether and C–C bonds are normally stable in these conditions, which mean that they usually remain unaltered during alkali treatment. During alkaline treatment of biomass, a great variety of compounds can be formed from lignin. Most of them are aromatic acids, aromatic aldehydes, and phenolic compounds that can have an inhibitory affect on enzymes and microorganisms.

5.3.3 Reactions between Ammonia and Lignocellulosic Biomass

In the special case of AFEX pretreatment of lignocellulosic biomass, ammonia is not only a source of hydroxyl groups but also a reactant. Amidation by ammonolysis of organic compounds occurs during this process and considerable amounts of acetamide were found in AFEX-treated corn stover water extracts.[11] The amidation agent in AFEX may be NH_3 in aqueous solution or in the vapor phase. The dissociation of ammonia in aqueous solutions and the dissociation of water produce hydroxyl ions that are responsible for the formation of hydroxy compounds. These reactions compete with some NH_3 for reactions that occur in the biomass during pretreatment. Ester bonds, for example, exist in certain monocot species in relatively large amounts and are one of the most reactive bonds in the presence of hydroxyl ions and ammonia.

The effects of aqueous ammonia solutions on different types of esters have also been reported.[8] It was found that the ratio of ammonolysis to hydrolysis decreases with increasing molecular weight of the alcohols associated with the different esters. Moreover, the reactivity decreases with the increase in the molecular weight of the ester. This is especially true in cases where the alcohol associated with the ester has side chains close to the ester bond (e.g. isopropyl acetate). The justification for this fact should lay on the stereochemistry of the chemical reaction and the size of ammonia compared to the hydroxyl ion.

Depending on the type of reaction, the resulting products will be either a carboxylic acid or an amide. In the case of the acetyl groups present in xylans and mannans, these reactions may produce acetic acid or acetamide depending if they are engaged in a hydrolysis or an ammonolysis reaction, respectively. Ammonolytic reactions target not only ester bonds, but also react with phenolic compounds, alkyl halogens, and alcohols.

5.3.4 AFEX Degradation Products

From our current understanding of AFEX pretreatment of biomass, there are a series of chemical and physical changes that occur which potentially improve enzyme accessibility to the substrate.[11] These changes result in increased digestibility when compared to untreated biomass. Some of the degradation products produced, such as aliphatic acids, furans, aromatic acids, and oligosaccharides in AFEX-pretreated poplar, have been published recently.[38] Other work has been performed to identify and quantify the degradation products that are formed during AFEX. In this study, corn stover was pretreated using AFEX at 100 °C, with a biomass to ammonia ratio of 1:1, 60% moisture, and 30 minutes of residence time. The degradation products were extracted from AFEX-treated corn stover with water and analyzed using LC-UV and LC-MS/MS.[11]

5.3.5 Waste Streams and Environmental Issues

When analyzing waste streams and environmental issues associated to a pre-treatment technology it is required to place the boundaries of the study around the whole biorefinery process. The reason for this fact is that the pretreatment choice will have impact on all the other unit operations, both upstream and downstream of pretreatment.

AFEX is known to be a dry-to-dry process, where both untreated and pretreated biomass enters and leaves the reactor in the dry form. The ammonia is evaporated and recycled to be reused in the process. In addition, AFEX-pretreated biomass does not need detoxification and nutrient supplementation to be hydrolyzed and fermented by enzymes and microorganisms.

In contrast to AFEX, other pretreatment processes often use larger amounts of water, producing solid and a liquid product streams. The solid stream (rich in cellulose) normally requires extensive washing steps before it can be hydrolyzed and fermented, which generates an important waste stream that needs to be taken into account for environmental and economical analysis. Both the solid and liquid products (rich in lignin and hemicellulose sugars) often require a detoxification step (e.g. overliming) that generates a solid waste (e.g. $CaSO_4$) which needs to be disposed or recycled.

Another important environmental factor to take into consideration is the carbon footprint of the catalyst. AFEX uses ammonia, which is currently produced by the Haber-Bosch process using natural gas as the hydrogen source. Hydrogen can be obtained using other sources of energy (e.g. solar), however, leaving ammonia with a minimal carbon footprint. The fact that ammonia is a gas at atmospheric pressure and at room temperature can raise some concerns about its utilization. In the biorefinery setting, ammonia will be used in large quantities and in a concentrated form, requiring important safety measures and control.

5.4 Enzymatic Hydrolysis

For laboratory screening purposes, enzymatic hydrolysis was performed at 1% glucan loading using 15 FPU of commercial cellulase per gram of glucan at pH 4.8 for a period of one week.[39] The hydrolysates were sampled at regular intervals to determine the sugar concentration using HPLC or enzyme-based sugar assays. The percentage glucan and xylan conversion helps us to understand the effectiveness of pretreatment for a given feedstock.[14] Enzyme cost is considered as a bottleneck to producing biofuels from lignocellulosic biomass. Several research groups are focusing on different strategies to find novel enzymes with high specific activity. Given the number of new commercial/novel enzymes and the number of different feedstocks with varying pretreatment conditions, conventional enzymatic hydrolysis screening strategies are tedious and time-consuming. To overcome this problem, highly automated, rapid assays using microplates to screen for optimal enzyme combinations have been developed.[40] Several issues have been addressed regarding solid delivery to microplates including transfer of biomass slurries, effect of particle size and mass transfer related parameters, optimal microplate methodology, reproducibility, and validity of the method compared to the conventional NREL protocol.[39] High solid loading hydrolysis (18%) is performed in order to obtain higher sugar concentrations needed for fermentation experiments. These high solid loading experiments must be performed using a fed-batch method, where biomass and enzymes are added at different intervals during hydrolysis. Unfortunately, the glucan and xylan conversions are significantly reduced relative to low solid loading experiments, primarily due to the presence of a higher concentration of sugars (substrate inhibition) and degradation products produced during pretreatment.[41] This problem can be overcome by using a simultaneous saccharification and fermentation (SSF) process, where the sugars produced during hydrolysis are simultaneously consumed by the microbes to produce alcohol.

5.5 Biomass Composition and Plant Species Classification

Bioenergy plants and model species can be grouped into a number of categories based on a system of plant classification (Figure 5.4A). Most energy species fall within two of the higher plant categories, the angiosperms (flower-bearing species) and the gymnosperms (cone-bearing species). The gymnosperms include all of the softwood species including pine, spruce, fir, and cedar. The angiosperms can be further sub-classified as monocots or dicots. The dicots include herbaceous flowering plants (e.g. alfalfa, soybean, tobacco) and all hardwoods (e.g. poplar, willow, black locust). The monocots include true grasses (e.g. switchgrass, orchardgrass, *Miscanthus*, sorghum, sugarcane, bamboo, etc.), grass-like species (sedges, rushes, cattails) and other herbaceous species.

Different plant species have different cell wall compositions (relative amounts of lignin, hemicellulose, cellulose, and ash) and different structures (macroscopic, microscopic, nanostructure, and molecular scales) within the cell wall. These factors are dependent on the cell wall chemistry, which may be similar within related species of plants. The most abundant cell wall polysaccharide is cellulose, making up 30–45% of the cell wall. The hemicelluloses, a class of polysaccharide, make up the next most abundant fraction (20–35%) of the walls. The most common hemicelluloses are glucouronoarabinoxylans, galactogluco- and glucomannans, xyloglucans, and mixed-linkage glucans.[42] The presence and relative amounts of each of these hemicelluloses in the plant wall depends greatly on the plant species. Dicots and some monocots have Type I primary cell walls, consisting of a network of cellulose microfibrils cross-linked with xyloglucans. This network is embedded in a complex matrix of pectins and proteins. The dicot secondary cell wall contains mainly xylan with little arabinosylation and a small proportion of mannans. All the grasses and some related monocots have Type II primary cell walls consisting of

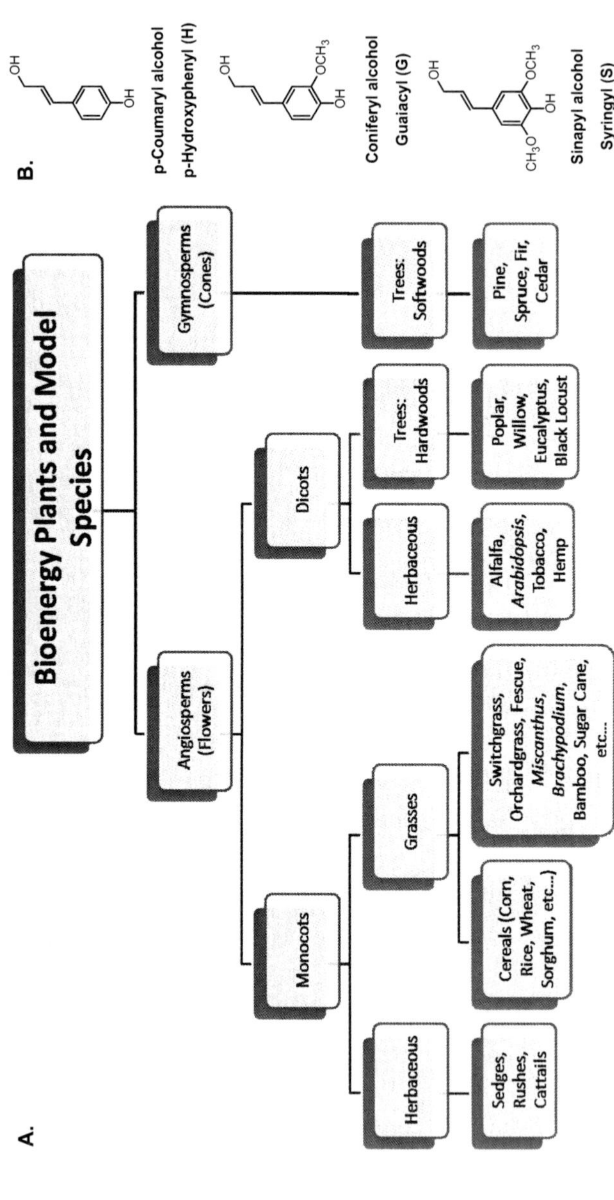

Figure 5.4 (A) Biomass plant species classification and (B) lignin building blocks in plants.

glucuronoarabinoxylans, the major cross-linking glycans, which are hydrogen bonded to the surfaces of cellulose microfibrils. The degree of arabinosylation can vary greatly between species. For example, wheat straw contains the lowest degree of substitution, whereas sorghum xylans have an exceptionally high degree of substitution. Gymnosperm secondary cell walls contain primarily mannans, such as *O*-acetylated galactoglucomannans, although ample amounts of xylans are also present.[43]

Lignin, a phenolic polymer, comprises 13–28% of the cell wall. There are three main monolignols that are incorporated into the larger, amorphous polymer: syringyl (S), guaiacyl (G), and p-hydroxyphenyl (H) (Figure 5.4B). Similarly to the hemicelluloses, the overall lignin content and composition can vary greatly between the different plant classes. The gymnosperms tend to have the highest relative amounts of lignin, which is composed mainly of G- and some H-units. The hardwoods and related herbaceous dicots have mainly G- and S-units, with some H-units. Grasses and related monocots have similar amounts of G- and S-units and significantly higher amounts of H-units than the dicots or gymnosperms.[44] In grasses, the lignin polymers are ester-linked to hemicellulose either through feruloyl ester linkages or through arabinose residues. Some of the hemicellulose residues are interlinked by diferulate bridges (Figure 5.5A).

These compositional and structural differences between various plant species can greatly impact a given pretreatment.[43] For example, ammonia is known to cleave ester linkages during pretreatment, which aides the release of lignin-oligosaccharides. So plant materials, such as grasses, which contain a high amount of glucuronoarabinoxylan ester linkages often respond very well to ammonia pretreatment.

5.6 AFEX Performance on Grasses

Because of the unique nature of the grass cell wall, grasses have generally been found to be highly digestible (>90% glucan and >70% xylan) after AFEX pretreatment using commercial cellulases and hemicellulases. The optimal AFEX conditions and glucan/xylan conversion for a variety of pretreated grasses are summarized in Table 5.2.

5.6.1 AFEX on Corn Stover

Corn stover is the most abundant agricultural residue produced in the Midwest of the USA, making it a prospective substrate for bioethanol production. Optimal conditions for AFEX pretreatment of corn stover were found to be 90 °C, 1:1 ammonia to biomass loading, 60% moisture, and 5 min residence time (when the set temperature is reached), based on maximum sugar yields for pretreated material by enzymatic hydrolysis.[45] The effect of grinding corn stover and fractionating it into various particle size ranges (untreated/AFEX-treated) on its composition and rate of enzymatic hydrolysis as well as the effect

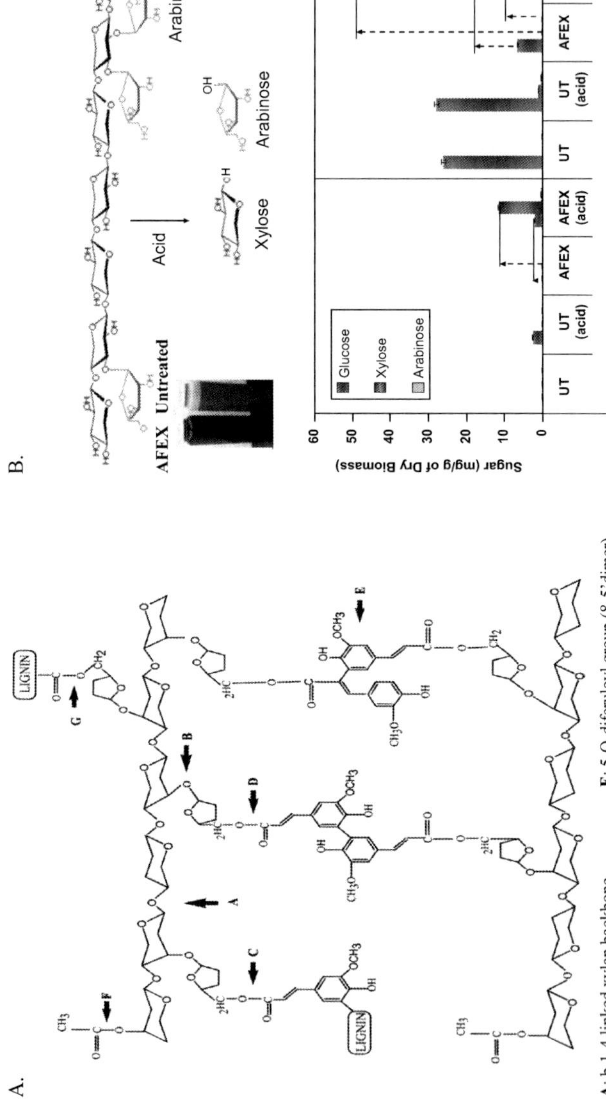

A: b 1-4 linked xylan backbone,
B: Xylose-arabinose linkage,
C: 5-O-feruloyl lignin,
D: 5-O-deferuloyl group (5-5'linked dimer),
E: 5-O-diferuloyl group (8-5' dimer),
F: 3-O-acetyl group,
G: arabinose-lignin

Figure 5.5 Pictures showing the simplified structure of arabinoxylan and how it can be quantified using acid hydrolysis: (A) arabinoxylan structure which is the main component of many hemicelluloses from primary cell walls of monocotyledonous plants with various ester and ether linkages; (B) how the broken xylo-oligosaccharides can be converted to monomeric sugars and quantified using HPLC.

Table 5.2 Biomass pretreatment conditions and respective glucan and xylan conversion for a given enzyme loading.

Biomass	Particle size	Pretreatment conditions	Glucan conversion (168 h)	Xylan conversion (168 h)	Glucan conversion (168 h)	Xylan conversion (168 h)	Enzyme loading/g of glucan	Reference
Sugar cane bagasse	10 mm screen	140 °C, 2:1, 150%, 30 min	20%	18%	90%	65%	33 mg of cellulose + 33 mg β-G + 8 mg of xylanase	52
Cane leaf	10 mm screen	100 °C, 1:1 60%, 30 min	32%	15%	90%	95%	33 mg of cellulose + 33 mg β-G + 15 mg of xylanase	52
Switchgrass	10 mm screen	100 °C, 1:1, 80%, 5 min	16%	3%	93%	70%	33 mg cellulase + 22 mg β-G	47
Reed canary grass vegetative stage	200 μm	100 °C, 1.2:1, 60%, 30 min	20%	0%	86%	78%	33 mg of cellulase + 33 mg β-G + 8 mg of xylanase	57
seed stage	200 μm	100 °C, 0.8:1, 60%, 30 min	20%	0%	89%	81%	33 mg of cellulase + 33 mg β-G + 8 mg of xylanase	57
Miscanthus	10 mm screen	140 °C, 2:1, 233%, 30 min	10%	6%	84%	85%	33 mg of cellulase + 33 mg β-G + 6 mg of xylanase	56
Rice straw	10 mm screen	140 °C, 1:1, 80%, 30 min	20%	10%	85%	86%	33 mg of cellulase + 33 mg β-G + 2.7 mg of xylanase + 3.7 mg pectinase	41, 49
Sorghum forage	10 mm screen	140 °C, 2:1, 120%, 5 min	22%	13%	90%	91%	33 mg of cellulose + 17 mg β-G + 15.6 mg of xylanase	54
sweet	10 mm screen	140 °C, 2:1, 140%, 30 min	10%	7%	81%	90%	33 mg of cellulose + 17 mg β-G + 15.6 mg of xylanase	54
Corn stover	10 mm screen	90 °C, 1:1, 60%, 5 min	32%	10%	95%	70%	33 mg of cellulose + 33 mg β-G + 3.3 mg of xylanase	14, 45
Elephant grass[*]	20 mm screen	90 °C, 1:1, 60%, 5 min	18%	10%	83%	51%	cellulase 5 IU/g of dry matter + 5.7 U of cellobiase	60

Table 5.2 (*Continued*)

Biomass	Particle size	Pretreatment conditions	Glucan conversion (168 h)	Xylan conversion (168 h)	Glucan conversion (168 h)	Xylan conversion (168 h)	Enzyme loading/g of glucan	Reference
Bermuda grass[*]	20 mm screen	90 °C, 2:1, 30%, 30 min	20%	NA	90%	NA	cellulase 5 IU/g of dry matter + 528.4 U/g of cellobiase	59
Alfalfa[*]	20 mm screen	85 °C, 2:1, 30%, 5 min	38%	34%	68%	86%	cellulase 5 FPU/g of dry matter	58
DDGS	10 mm screen	90 °C, 1:1, 40%, 5 min	63%	6%	85%	81%	cellulase (15 FPU/g of cellulose) + β-G (40 IU/g of cellulose) + pectinase (50 U xylanase/g DDGS db) + feruloylesterase (2 UI/g of DDGS)	62
Hardwood poplar	10 mm screen	180 °C, 2:1, 233%, 30 min	5%	5%	93%	65%	125 mg cellulase + 33 mg βG + 125 mg of xylanase	38

NA – not available
*hydrolysis experiments done at 24 h

of water washing the AFEX-treated fractions on enzymatic hydrolysis have been explored.[14] Particle size reduction and washing were found to improve effectiveness of AFEX pretreatment and substantially improve the hydrolysis yields. The time required for complete glucan hydrolysis of milled and washed AFEX corn stover (supplemented by commercially available xylanase) was reduced by 96 h compared to the unwashed samples (168 h).[14]

5.6.2   AFEX on Switchgrass

Switchgrass is a hardy, deep-rooted, perennial, warm season grass native to North America. There are two ecotypes of switchgrass in the USA: an upland ecotype, often found in the northern USA, favoring drier soils, and faring better in semi-arid climates; and a lowland ecotype, most suited to the southern USA, favoring heavier soils, and found where water availability is more reliable. The upland ecotype includes the varieties Cave-in-Rock, Trailblazer, Pathfinder, Blackwell, and Caddo. The lowland ecotype includes such varieties as Alamo and Kanlow.[46] In one study, the optimal pretreatment conditions (100 °C, 1:1 ammonia to biomass loading, 80% moisture, and 5 min) were selected based on maximum glucan (93%) and xylan (70%) conversion to fermentable sugars using a fixed amount of enzyme when compared to 16% and 3% for untreated samples, respectively.[47] Harvest time greatly influences the composition and pretreatment severity. The early harvest samples have lower glucan and lignin contents and a higher protein content and also required lower temperatures and less ammonia during pretreatment when compared to late harvest samples.

5.6.3   AFEX on Rice Straw

With nearly 800 million dry tons produced worldwide annually, rice straw (RS) holds great potential for the cellulosic ethanol market. It has a higher ash content (mostly silica) compared to other grasses. RS was found to be more recalcitrant than other agricultural residues such as corn stover.[48] Therefore, a mushroom spent straw (MSS)/AFEX technology integrating a biological pre-treatment by white rot fungi followed by AFEX pretreatment was used.[49] Treating RS with a white rot fungus (*Pleurotus ostreeatus*) benefits enzymatic hydrolysis by selectively degrading recalcitrant lignin, while the fungus utilizes free sugars and co-hydrolyzed oligosaccharides for growth of their fruiting bodies.[49] Optimal AFEX pretreatment conditions for rice straw were found to be 140 °C, 1:1 ammonia to biomass loading, 80% moisture, and 30 min. The highest glucan conversion for 1% solid loading was found to be 85%. Further, the xylan conversion was improved from 55% to 82% by appropriate addition of Multifect® Xylanase and Multifect® Pectinase in addition to Spezyme CP cellulase.[41] A large amount of oligomers were still present in the hydrolyzate after hydrolysis, and as the solid loading was increased from 3% to 18%, the concentration of oligosaccharides also increased.

5.6.4 AFEX on Sugarcane Bagasse

In 2005, 1324.6 million tons of sugarcane were grown in tropical regions such as Brazil, India, Cuba, and China for the commercial production of sugar.[50] For every 10 tonnes of sugarcane crushed, a sugar factory produces nearly 3 tonnes of wet bagasse. About half of the bagasse produced in a sugar mill is burned to generate energy to support the sugar refining process. The remaining bagasse could potentially be used utilized for bioethanol production following pre-treatment, enzymatic hydrolysis, and fermentation. There are very few reports on pretreating sugarcane bagasse using ammonia[51,52] to produce biofuels. Recent findings at Michigan State University showed very promising conversion for AFEX-treated bagasse and cane leaf matter. The optimal AFEX conditions for bagasse were found to be 140 °C, 2:1 ammonia to biomass loading, 140% moisture, and 30 min residence time. For cane leaf matter the optimal conditions were found to be 100 °C, 1:1 ammonia to biomass loading, 60% moisture, and 30 min residence time. The glucan conversion for both these materials was found to be close to 90%, while the xylan conversions were found to be 65% and 95%, respectively.

5.6.5 AFEX on Sorghum

Based on a recent economic analysis, sweet sorghum is considered to be one of the most drought-resistant crops available and has higher biomass yield and lower production costs than many other plants.[53] Forage sorghum, which is typically grown for animal feed as silage, has a very short growing period (about 60 days), much shorter than the growing period required for corn or other kinds of sorghum (no less than 120 days) including grain sorghum and sweet sorghum. AFEX pretreatment conditions were optimized for both sweet sorghum and forage sorghum biomass. Supplementing xylanase with cellulase during enzymatic hydrolysis of AFEX-pretreated sorghum increased both glucan and xylan conversion to 90% at 1% glucan loading.[54] Sorghum biomass was further hydrolyzed at high solid loadings and then fermented to ethanol using *Saccharomyces cerevisiae* 424A (LNH-ST) without any external nutrient supplementation or detoxification. A complete mass balance for the pretreatment, hydrolysis, and fermentation process is summarized in the manuscript.[54]

5.6.6 AFEX on *Miscanthus*

Miscanthus (*Miscanthus x giganteus*) can grow over 12 feet tall and has attracted considerable attention in Europe and the USA as a possible dedicated energy crop, either as fuel for electricity generation or, more recently, for conversion to a biofuel such as ethanol. Numerous agricultural studies throughout Europe have been promising, showing yields ranging from 25 t/acre in Britain and Denmark to 30 t/acre in Spain and Italy.[55] The optimal AFEX conditions for Miscanthus were found to be 140 °C, 2:1 ammonia to biomass loading, 233% moisture, and 30 min residence time. Glucan and xylan conversions close to 90–95% and 80–85%, respectively, were possible upon hydrolysis of AFEX-treated *Miscanthus* using cellulases (15 FPU/(g of glucan)) and other additives, such as xylanase (0.53 g/(100 g of *Miscanthus*)) or pectinase (1.03 g/(100 g of *Miscanthus*)) and Tween-80 (0.35 g/(g of glucan)).[56]

5.6.7　AFEX on other Grasses and Biomass

Other feedstocks which were pretreated using AFEX subsequently hydrolyzed using commercial enzymes include reed canarygrass,[57] alfalfa,[58] bermuda grass,[59] elephant grass,[60] a number of legumes, and distiller's dry grains with solubles (DDGS). DDGS is the major co-product in corn ethanol plants, and is sold as an animal feed. The increase in DDGS supply due to increased ethanol production is expected to drive its value down, and thus further value addition is necessary. One approach is to convert the cellulose in DDGS into sugars for increased ethanol production, leaving a residue higher in protein content. AFEX pretreatment followed by enzymatic hydrolysis of both wet and dry DDGS was studied. Optimal AFEX conditions for dry and wet distiller's grains were 70 °C and 0.8 kg anhydrous NH_3/kg dry biomass and 80 °C and 0.6 kg NH_3/kg dry biomass, respectively. Xylose yields were negligible.[61] Xylose conversion was a major concern during enzyme hydrolysis, and subsequent enzyme optimization using different hemicellulases significantly improved xylan conversion from 14% to 81% conversion.[62] Subsequently, the pretreated DDGS was hydrolyzed at high solid loadings and found to be highly fermentable to ethanol.[63]

5.7　AFEX Comparison on Grasses versus Hardwoods

When compared to grasses (monocots), hardwood (dicots) plant species have a significantly different complex cell wall structure, as explained earlier in this chapter. Hardwoods (e.g. poplar) are a good source of cellulosic fiber, and their lignin and monolignol composition are very different from grasses (e.g. corn stover). Poplar requires much higher temperature, moisture, and ammonia loadings during AFEX pretreatment to achieve significant glucan hydrolysis yields when compared with corn stover. Even at higher temperatures (e.g. 180 °C), yields only reached to 50% glucan and 35% xylan conversion with 1:1 ammonia to biomass loadings for poplar. On the other hand, for corn stover, the best AFEX condition was found to be close to 90 °C, 60% moisture, and 1:1 ammonia to biomass loadings.

Most of the arabinoxylan in the cell wall is connected to lignin through ether and ester linkages. These linkages are typically between glucuronic acid and arabinose side chains of hemicelluloses and hydroxyl/carboxyl functionalities of lignin (e.g. ferulic and coumaric acid) (Figure 5.5A). Ammonia has a tendency to cleave these ester linkages via ammonolysis. The arabinoxylan content of grasses (3–6%) is significantly higher than hardwoods (<1%, for poplar). With a higher arabinoxylan content of the cell wall, more ester linkages are likely to be cleaved during AFEX pretreatment. Arabinoxylans help form bridges between lignin and hemicellulose/cellulose that reduce enzyme accessibility. Cleavage of these linkages would help increase the cell wall pore volume, reduce the protective barrier of lignin, and enhance enzyme accessibility to cellulose and hemicellulose. This is one factor that could explain why grasses are more easily digestible after AFEX pretreatment when compared with hardwood poplar. Further support for this hypothesis would come from detailed quantification of oligosaccharides produced during AFEX pretreatment for both corn stover and poplar (Figure 5.5B). The wash streams generated from untreated/AFEX-treated poplar and corn stover were analyzed for degradation products, monomeric and oligomeric sugars.[38] AFEX pretreated poplar wash stream contained very low concentrations of monomeric sugars prior to acid hydrolysis. After acid hydrolysis we observe a small increase in glucose concentration, a 54-fold increase in xylose concentration and a 0.8-fold increase in arabinose concentration (Figure 5.5B). Upon performing mild acid hydrolysis of AFEX-treated corn stover wash stream, we observe about 2.9-, 300-, and 79-fold increase in glucose, xylose, and arabinose concentrations, respectively.

5.8   Advantages of AFEX during Fermentation

AFEX-treated corn stover hydrolyzate with lower amounts of degradation products were found to be highly fermentable.[64] The residual ammonia left in the biomass can be utilized by microbes during the fermentation process. Fermentations using *S. cerevisiae* 424A(LNH-ST) in complex media containing YEP and AFEX-corn stover hydrolysates (6% glucan loading/18% solids loading) were compared under identical conditions.[65] These hydrolysates were fermented without conditioning (no washing, nutrient supplementation, or detoxification) at 1.1 g dry-cell-wt./L starting cell density.[41,54,64] All of the above-mentioned steps are associated with the cost in a biorefinery and by using AFEX-treated corn stover, we can have a substantial cost and energy savings by avoiding such steps. Strain 424A(LNH-ST) was shown to grow well in both complex media and AFEX-CS hydrolysates and achieved cell densities greater than 6.0 g/L within 12 h of fermentation. For fermentations in AFEX corn stover hydrolysate, volumetric xylose consumption rates were an order of magnitude lower than for glucose. This showed that xylose utilization was highly susceptible to inhibition compared to glucose utilization. Surprisingly, specific glucose consumption rates in hydrolysate achieved 10.6 g/L/hr/g cell; 0–6 h, substantially higher than that in complex media.[65]

5.9   Logistics and Regional Biomass Processing Centers

The low bulk density of lignocellulosic material combined with the large (>1000 tons per day) refinery sizes required to achieve reasonable economies of scale can cause transportation and logistical problems that must be solved.

In order to supply the refinery, a company must contract with hundreds or thousands of farmers to obtain enough feedstock, another logistical issue. A final drawback is that, while the capital cost per gallon of ethanol decreases with increasing capacity of the refinery, the overall capital costs will still increase. This capital cost is quite large[66] and thus incurs a great deal of risk to build a new facility.

One approach to reducing or eliminating these hurdles is to create Regional Biomass Processing Centers (RBPCs).[67] These centers would collect biomass from the surrounding area and process it into a form suitable for a biorefinery. Several smaller RBPCs could surround and supply one large refinery. In the simplest form, this would merely involve storage of biomass and possibly grinding or bailing the biomass. Other value-added products such as proteins, lipids, or specialty chemicals could be extracted or produced here if no integration with lignocellulosic refining is required. Finally, pretreatment itself could be decoupled from the biorefinery and performed at the RBPC, particularly if AFEX treatment is used.

5.10 Pellets and Logistics of Transportation

Densification of lignocellulosic biomass has been considered as a method to reduce transportation costs to the refinery. After grinding the biomass, it can be densified to pellets, cubes, or briquettes, increasing the density from 70 to 450–650 kg/m^3.[68] This can reduce the number of trucks required at the biorefinery, which may be a necessity given road congestion and unloading logistics. In addition, there is added benefit in reducing the cost of diesel. However, the price of pelletization can be inhibitory to the process. It has been estimated cost up to $30 per ton to pelletize biomass.[68] Of that cost, $10/ton is drying the biomass, which may not be an issue if harvested at a late enough date. AFEX-pretreated material may reduce this cost, however. The lignin deposited on the surface of the biomass during AFEX pretreatment acts as a natural binding agent, thereby decreasing the pressure required to densify the material. By pretreating the material first, the price for densification may be decreased to less than $10/ton, making it economically competitive with delivering bulk material to the refinery. As the size of refineries increase (>5000 tons per day), densification at a RBPC becomes more economically attractive.

5.11 Storage and Stability

As explained in the fundamental understanding of the AFEX process, ammonia cleaves the ester linkages which aid in the solubilization of lignin in ammonia. After the pretreatment when the reactor is depressurized, the liquid ammonia exits the reactor as gaseous ammonia, leaving behind the solubilized lignin and oligosaccharides on the surface of the biomass. This lignin on the surface of the biomass acts as a preservative and pretreated biomass is quite stable for months if stored with less then 10% moisture at room temperature in dry conditions. Further, there is no separate liquid stream generated during the AFEX pretreatment process and all the minerals and nutrients (proteins, amino acids, and vitamins) remain in the biomass.

5.12 Co-producing Animal Feeds and Biofuels using AFEX Pretreatment

Our results showed that AFEX pretreatment improved the digestibility of neutral detergent fiber (NDF) for multiple feedstock's during *in vitro* rumen fermentation studies.[69] NDF is a common measure of fiber for feeds, and its digestibility strongly influences the overall value of the material. Common feeds, such as alfalfa, corn silage, or orchardgrass, did not see a large improvement in digestibility when AFEX-treated. Instead, the greatest improvements were in moderately indigestible material not commonly used as cattle feed. Of particular interest is corn stover and late-harvest switchgrass, which saw 52% and 128% improvement in 48-h digestibility over untreated material and 74% and 70% improvement over samples that have been treated using conventional ammoniation techniques, respectively (unreported results). AFEX treatment improved corn stover's total digestible nutrients to a level comparable with highly digestible fiber sources such as soybean hulls, while improving switchgrass's nutrient value to levels comparable to traditional forages.

Preliminary analysis suggests that this approach can dramatically reduce the amount of land needed for current food production, and therefore increase the potential biofuel capacity. Switchgrass can produce approximately twice as much digestible fiber per acre as alfalfa and approximately four times as much as orchardgrass. In addition, there is more crude protein from AFEX-treated switchgrass per acre than these forages (unreported results). Additionally, these forages, particularly alfalfa, can produce more protein per acre than soybeans while simultaneously producing cellulosic ethanol. Cover crops can also be grown on the same land as corn grain and provide additional protein and ethanol. By combining these advantages, we predict that approximately 65% of US gasoline demand can be displaced by biofuels grown on farmland without harming food production.

5.13 Economic Considerations

The primary purpose of pretreatment is to make the subsequent hydrolysis and fermentation of lignocellulosic sugars economically viable. According to a study by NREL, pretreatment costs can account for over 20% of the capital costs in the biorefinery.[66] Compared to other pretreatments, AFEX has slightly higher capital and utility costs, but low material costs. However, overall sugar yields dominate the economics of any pretreatment process, and the high yields produced by AFEX allows it to be an economically attractive pretreatment choice.[6] Much of the capital and utility cost of AFEX is due to the need to recover the ammonia, and so this area has been the focus of AFEX economic modeling.

While most ($\sim$90%) of the ammonia will vaporize when the pressure is released, the rest remains soluble in the water present within the biomass. The basic design[6,70] requires a dryer and distillation after the initial flash. The dryer removes the residual ammonia as well as some water, which are separated in the distillation column. Due to the high capital cost of this process, the ammonia recovery must be carefully designed to reduce the costs. Of particular interest is the dryer, which should be designed to remove the ammonia with as short a residence time as possible (thereby reducing capital costs) and by minimizing the amount of water that vaporizes (thereby reducing the size and duty on the distillation column). Carefully controlling the temperature and vacuum pressure is critical for this design. In addition, the use of sweep gases, including nitrogen or supercritical ammonia, can also improve the process. In the case of nitrogen, a separate compressor or condenser must be used to remove the ammonia from the nitrogen prior to the distillation column. Recent results indicate that concentrated aqueous ammonia can be fed to the biomass without any reduction in AFEX treatment quality[70] which could eliminate the need for a distillation column. Likewise, if the ammonia can be added to the biomass in the gas phase, then the condenser can also be eliminated.

5.14 Conclusions

A comprehensive summary of alkaline pretreatment, and more specifically AFEX pretreatment, has been given. The advantages and disadvantages have been discussed in detail in this review. The fundamental understanding of how AFEX pretreatment helps to deconstruct the biomass will help to further improve the process in the future.

Acknowledgements

This work was funded by DOE Great Lakes Bioenergy Research Center (www.greatlakesbioenergy.org) supported by the U.S. Department of Energy, Office of Science, Office of Biological and Environmental Research, through Cooperative Agreement DE-FC02-07ER64494 between the Board of Regents of the University of Wisconsin System and the U.S. Department of Energy. We also appreciate financial support, in initial stages of the project, from Michigan State Research Foundation (SPG grant). We would like to thank James Humpula and Shishir Chundawat for helping us to edit and revise the manuscript.

References

1. H. E. Grethlein, *Biotechnol. Adv.*, 1984, **2**, 43.
2. C. W. Wyman, B. E. Dale, R. T. Elander, M. Holtzapple, M. R. Ladisch and Y. Y. Lee, *Bioresour. Technol.*, 2005, **96**, 1959.

3. R. P. Chandra, R. Bura, W. E. Mabee, A. Berlin, X. Pan and J. N. Saddler, *Adv. Biochem. Eng. Biotechnol.*, 2007, **108**, 67.

4. A. T. Hendriks and G. Zeeman, *Bioresour. Technol.*, 2009, **100**, 10.

5. L. da Costa Sousa, S. P. Chundawat, V. Balan and B. E. Dale, *Curr. Opin. Biotechnol.*, 2009, **20**, 339.

6. T. Eggeman and R. Elander, *Bioresour. Technol.*, 2005, **96**, 2019.

7. D. Perkis, W. Tyner and R. Dale, *Bioresour. Technol.*, 2008, **99**, 5243.

8. S. P. S. Chundawat, R. Vismeh, L. N. Sharma, J. F. Humpula, L. da Costa Sousa, C. K. Chambliss, A. D. Jones, V. Balan and B. E. Dale, *Bioresour. Technol.*, 2010, **101**, 8429.

9. E. Sjöström, Tappi, 60, 1977, **9**, 151.

10. D. Fengel and G. Wengener, *Wood. Chemistry, Ultrastructure, Reactions*, Walter de Gruyter, Berlin, Germany, p. 613, 1989.

11. S. P. S. Chundawat, B. S. Donohoe, L. Dacosta Sousa, T. Elder, U. P. Agarwal, F. Lu, J. Rolph, M. E. Himmel, V. Balan and B. E. Dale, *Energy Environ. Sci.*, 2011, **4**, 973.

12. S. G. Pavlostathis and J. M. Gossett, *Biotechnol. Bioeng.*, 1985, **27**, 334.

13. D. Gregg and J. N. Saddler, *Appl. Biochem. Biotechnol.*, 1996, **711**.

14. S. P. Chundawat, V. Balan and B. E. Dale, *Biotechnol. Bioeng.*, 2007, **96**, 219.

15. R. H. Marchessault and P. R. Sundararajan, *Cellulose*, in *The Polysaccharides*, ed. G. O. Aspinall, Academic Press, Inc., New York, 1983, Vol. 2, pp. 11–95.

16. R. C. Pettersen, *The chemical composition of wood (chapter 2)*, in *The Chemistry of Solid Wood, Advances in Chemistry Series*, ed. R. M. Rowell, American Chemical Society, Washington, DC, 1984, Vol. 207, p. 984.

17. A. Sarko, *Recent X-ray crystallographic studies of celluloses*, in *Cellulose: Structure, Modification, and Hydrolysis*, eds. R. A. Young and R. M. Rowell, Wiley-Interscience, New York, 1986.

18. M. Wada, L. Heux, Y. Nishiyama and P. Langan, *Biomacromolecules*, 2009, **10**, 302.

19. R. G. Zhbankova, S. P. Firsova, D. D. Grinshpanb, J. Baranc, M. K. Marchewkac and H. Ratajczak, *J. Mol. Structure*, 2003, **645**, 9.

20. P. J. Weimer, A. D. French and T. A. Calamari, *Appl. Environ. Microbiol.*, 1991, **57**, 3101.

21. K. Igarashi, M. Wada and M. Samejima, *FEBS J.*, 2007, **274**, 1785.

22. R. Sierra, C. B. Granda and H. T. Holtzapple, *Methods Mol. Biol.*, 2009, **581**, 115.

23. T. H. Kim, R. Gupta and Y. Y. Lee, *Methods Mol. Biol.*, 2009, **581**, 79.

24. V. Balan, B. Bals, S. P. Chundawat, D. Marshall and B. E. Dale, *Methods Mol. Biol.*, 2009, **581**, 61.

25. P. J. Weimer, Y.-C. T. Chou, W. M. Weston and D. B. Chase, *Biotechnol. Bioeng. Symp.*, 1986, **17**, 5.

26. T. C. Yu-Chia, *Biotechnol. Bioeng. Symp.*, 1987, **17**, 19.

27. S. B. Kim and Y. Y. Lee, *Appl. Biochem. Biotechnol.*, 1996, **57/58**, 147.

28. J. Gould, *Biotech. Bioeng.*, 1984, **26**, 46.

29. L. C. Teixeira, J. C. Linden and H. A. Schroeder, *Appl. Biochem. Biotechnol.*, 1999, **77–79**, 19.
30. M. Appl, Ammonia, in *Ullmann's Encyclopedia of Industrial Chemistry*, Wiley-VCH, United States Geological Survey publication, Weinheim, 2006, http://minerals.usgs.gov/minerals/pubs/commodity/nitrogen/nitromcs 05.pdf. Retrieved 2009-07-07.
31. A. J. Barry and F. C. Peterson, *J. Am. Chem. Soc.*, 1936, **58**, 333.
32. J. J. O'Corner, *Tapii*, 1972, **55**, 353.
33. J. T. Huber, H. F. Bucholtz and R. L. Boman, *J. Dairy Sci.*, 1980, **63**, 76.
34. J. Zorrilla-Rios, G. W. Horn, W. A. Phillips and R. W. McNew, *J. Anim. Sci.*, 1991, **69**, 1809.
35. B. E. Dale and M. J. Moreira, Biotech. and Bioeng. Symp #12, '*Biotechnology in Energy Production and Conversation*', 1982, p. 13.
36. B. E. Dale, Method for increasing the reactivity and digestibility of cellulose with ammonia. US Patent No. 4600 590, 1986.
37. J. B. Dunson, M. P. Tucker, R. T. Elander and R. C. Lyons. System and process for biomass treatment. US. Pub. No. US2007/0029252A1, 2007.
38. V. Balan, L. da Costa Sousa, S. P. Chundawat, D. Marshall, L. N. Sharma, C. K. Chambliss and B. E. Dale, *Biotechnol. Prog.*, 2009, **25**, 365.
39. NREL LAP-000 protocol for hydrolysis of biomass. http://www.nrel.gov/ biomass/analytical_procedures.html
40. S. P. Chundawat, V. Balan and B. E. Dale, *Biotechnol. Bioeng.*, 2008, **99**, 1281.
41. C. Zhong, M. W. Lau, V. Balan, B. E. Dale and Y. J. Yuan, *Appl. Microbiol. Biotechnol.*, 2009, **84**, 667.
42. M. Pauly and K. Keegstra, *Plant J.*, 2008, **54**, 559.
43. M. C. McCann and N. C. Carpita, *Curr. Opin. Plant Biol.*, 2008, **11**, 314.
44. W. Boerjan, J. Ralph and M. Baucher, *Annu. Rev. Plant Biol.*, 2003, **54**, 519.
45. F. Teymouri, L. Laureano-Perez, H. Alizadeh and B. E. Dale, *Bioresour. Technol.*, 2005, **96**, 2014.
46. D. J. Parrish and J. H. Fike, *Methods Mol. Biol.*, 2009, **581**, 27.
47. H. Alizadeh, F. Teymouri, T. Gilbert and B. Dale, *Appl. Biochem. Biotechnol.*, 2005, **121–124**, 1133.
48. L. E. Gollapalli, B. E. Dale and D. M. Rivers, *Appl. Biochem. Biotech.*, 2002, **98–100**, 23.
49. V. Balan, L. da Costa Sousa, S. P. Chundawat, R. Vismeh, A. D. Jones and B. E. Dale, *J. Ind. Microbiol. Biotech.*, 2008, **35**, 293.
50. UNFAO (http://www.fao.org/es/ess/top/commodity.html?lang=en&item= 156&year=2005).
51. B. A. Prior and D. F. Day, *Appl. Biochem. Biotechnol.*, 2008, **146**, 151.
52. C. Krishnan, C. L. da Costa Sousa, M. Jin, L. Chang, B. E. Dale and V. Balan, *Biotechnol. Bioeng.*, 2010, **107**, 441.
53. D. Y. Corredor, J. M. Salazar, K. L. Hohn, S. Bean, B. Bean and D. Wang, *Appl. Biochem. Biotechnol.*, 2009, **158**, 164.
54. B. Z. Li, V. Balan, Y. J. Yuan and B. E. Dale, *Bioresour. Technol.*, 2010, **101**, 1285.

55. E. Heaton, T. Voigt and S. P. Long, *Biomass Bioenergy*, 2004, **27**, 21.
56. H. K. Murnen, V. Balan, S. P. Chundawat, B. Bals, L. da Costa Sousa and B. E. Dale, *Biotechnol. Prog.*, 2007, **23**, 846.
57. T. C. Bradshaw, H. Alizadeh, F. Teymouri, V. Balan and B. E. Dale, *Appl. Biochem. Biotechnol.*, 2007, **134–140**, 395.
58. A. Ferrer, F. M. Byers, B. Sulbaran de Ferrer, B. E. Dale and C. Aiello, *Appl. Biochem. Biotechnol.*, 2002, **98–100**, 123.
59. S. Reshamwala, B. T. Shawky and B. E. Dale, *Appl. Biochem. Biotechnol.*, 1995, **51/52**, 43.
60. A. Ferrer, F. M. Byers, B. Sulbaran-de-Ferrer, B. E. Dale and C. Aiello, *Appl. Biochem. Biotechnol.*, 2000, **84–86**, 163.
61. B. Bals, B. E. Dale and V. Balan, *Energy Fuels*, 2006, **20**, 2732.
62. B. S. Dien, E. A. Ximenes, P. J. O'Bryan, M. Moniruzzaman, X. L. Li, V. Balan, B. E. Dale and M. A. Cotta, *Bioresour. Technol.*, 2008, **99**, 5216.
63. Y. Kim, R. Hendrickson, N. S. Mosier, M. R. Ladisch, B. Bals, V. Balan and B. E. Dale, *Bioresour. Technol.*, 2008, **99**, 5206.
64. M. W. Lau, B. E. Dale and V. Balan, *Biotechnol. Bioeng.*, 2008, **99**, 529.
65. M. W. Lau and B. E. Dale, *Proc. Natl. Acad. Sci. USA*, 2009, **106**, 1368.
66. A. Aden, 2007 state of technology model, in: *N.R.E.L.T. Report* (Ed.), 2008.
67. J. E. Carolan, S. V. Joshi and B. E. Dale, *J. Agric. Food Ind. Org.*, 2007, **5**, 10.
68. S. Mani, S. Sokhansanj, X. Bi and A. F. Turhollow, *Appl. Eng. Agric.*, 2006, **22**, 421.
69. P. J. Weimer, D. R. Mertens, E. Ponnampalam, B. F. Severin and B. E. Dale, *Animal Feed Sci. Technol.*, 2003, **103**, 41.
70. E. Sendich, M. Laser, S. Kim, H. Alizadeh, L. Laureano-Perez, B. E. Dale and L. R. Lynd, *Bioresour. Technol.*, 2008, **99**, 8429.
71. R. A. Silverstein, Y. Chen, R. R. Sharma-Shivappa, M. D. Boyette and J. Osborne, *Bioresour. Technol.*, 2007, **98**, 3000.

Cellulases and Hemicellulases for Biomass Degradation: An Introduction

SUPRATIM DATTA[a] AND RAJAT SAPRA[b]

[a] Deconstruction Division, Joint BioEnergy Institute, 5885 Hollis St, 4th Floor, Emeryville, CA, USA; [b] Biomass Science and Conversion Technology Department, Sandia National Laboratory, Livermore, CA, USA

6.1 Introduction

In 2008, the United States consumed a fourth of the total world oil consumption and imported $\sim 57\%$ of its total petroleum needs making it the largest consumer of oil in the world.[1] The US National Energy Policy aims to decrease the dependence on foreign oil by increasing domestic energy supplies using a more diverse mix of energy resources. One of the sources for renewable, secure and domestic supply of energy is plant biomass, henceforth referred to as lignocellulosic biomass. Lignocellulose is a natural abundant material created by plants, from sunlight, soil nutrients and carbon dioxide in the atmosphere. Cellulose, hemicellulose and lignin are the primary plant biomass components that are used for energy production. Lignocellulosic biomass includes agricultural residues, deciduous and coniferous woods, forest, bioenergy crops and municipal solid wastes and wastes from pulp and paper industry. A study published by Oak Ridge National Laboratory in April 2005, found that the United States could potentially generate 30% of energy equivalent of the

RSC Energy and Environment Series No. 4
Chemical and Biochemical Catalysis for Next Generation Biofuels
Edited by Blake Simmons

Published by the Royal Society of Chemistry, www.rsc.org

petroleum consumption using lignocelluloses as the feedstock.[2] The main steps involved for production of fuels from lignocellulosic biomass involve feedstock preparation, pretreatment to reduce crystallinity of the biomass, fractionation of the pretreated biomass, enzyme production, hydrolysis of the biomass by enzymes to produce sugars, fermentation of the sugars to produce fuels, product recovery, and waste treatment.

However, there are certain technological and economic hurdles to overcome before ligocellulosic fuels can be cost competitive with fossil fuels. The high cost of enzymes required to hydrolyze biomass is one of the major hindrance to make the biofuel production cost comparable to current fossil fuel costs.[3] Most of available literature reports on high saccharification yields for pretreated biomass are based on high enzymes loading making the process commercially unfeasible.[4] The higher prices of enzymes historically were due to cost of growing *Trichoderma reesei* from which most industrial cellulases were derived and the low specific activity of the isolated enzyme. Even though a 10-fold cost reduction was achieved recently, cellulases are much more expensive than the starch degrading enzymes produced industrially.[5] To produce cellulases at comparable costs and generate lignocellulose derived sugars cheaply, better enzymes need to be engineered. To engineer efficient enzymes, a better understanding of structure and function of cellulases is required. This chapter introduces cellulases and hemicellulases that are produced by cellulolytic microorganisms which contain a wide variety of different catalytic and carbohydrate binding modules and act synergistically to degrade cellulose. In some microbes, cellulases are organized into an elaborate multifunctional supramolecular complex, known as the cellulosome. Insights into the structure of cellulases, hemicellulases and cellulosomes and the mechanisms of its hydrolysis are discussed.

6.2 Why is Lignocellulose so Hard to Break Down?

Plant cell walls are complex structures composed mostly of lignocellulose which is a matrix of cross-linked polysaccharide networks and lignin. In an oversimplified picture, the chemical cross-linkages, intermolecular bonds and the matrixing of the components lead to the recalcitrance of cellulose. Cellulose is the main component of plant cell walls and is highly crystalline and resistant to depolymerization. It is composed of 8,000–12,000 D-glucose units connected by $\beta(1-4)$ glycosidic bonds. The two ends of a cellulose chain are known as reducing and non-reducing ends. The reducing end is so called because the hemiacetal is able to open to expose the reducing aldehyde. On the non-reducing end, the first carbon in the hemiacetal is involved in the $\beta(1-4)$ bond which prevents ring opening (Figure 6.1). Thus cellulose chains are directional with cellobiose (glucose dimer) as the repeating unit. The glucose pyranose structure aligns itself alternatively because of the presence of the $\beta(1-4)$ linkage and the chains are arranged in layer sheets in cellulose crystals. This was experimentally confirmed by X-ray scattering patterns of pure crystalline cellulose from different sources. The chains within each sheet are linked

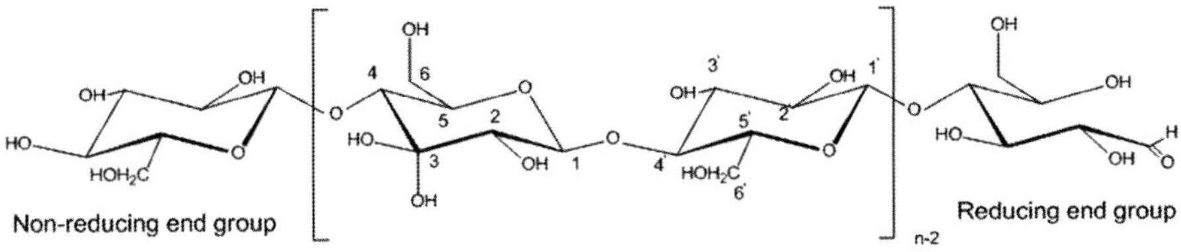

Figure 6.1 Cellulose chain is made up of a reducing end and a non-reducing end.

through hydrogen bonds and align in a parallel or antiparallel direction. Van der Waals forces help the sheet stack together.[6] Crystalline cellulose has been found to exist in different forms. I_α and I_β cellulose are the most common forms. Cell walls of algae and bacteria are rich in I_α while cotton, wood and ramie fibers are predominantly made up of I_β form. The other polymorphs cellulose II, III and IV are the result of cellulose pre-treatment with natural celluloses undergoing reversible and/or irreversible conversions. The structures of both forms of natural celluloses have been determined by Nishiyama *et al.*[7,8] I_α is made up of a triclinic P_1 unit cell made up of one chain that has two geometrically non-identical adjacent glucose residues that are related by a pseudo two-fold axis of symmetry while the I_β has a monoclinic $P2_1$ unit cell that contains two conformationally distinct glucose chains. The main difference in chain packing is the parallel chain packing arrangements for forms I and III (reducing ends at one end of the crystal) and antiparallel arrangement for cellulose II (alternate chains in opposite directions). As will be discussed later, the stacking of aromatic amino acid residues on the exposed rings of cellulose are responsible for the interaction between the carbohydrate binding domains of cellulases and the crystalline cellulose substrate.[9] These residues overlap with the pyranose rings of cellulose chains thereby increasing the potential for CH-π interactions.

6.3　Pretreatment of Cellulose

The pretreatment of lignocellulosic biomass is crucial for successful enzymatic conversion of biomass to fermentable sugars. Pretreatment options have to be severe enough to break apart the upper layers of the plant cell microfibrils but not enough to degrade the sugars to produce toxic compounds as effluents or toxic to the downstream processes. Pretreatment regimes are commonly used to fractionate, solubilize, hydrolyze and separate cellulose, hemicellulose and lignin components. Current pretreatment options used are mainly derived from the pulp and paper industry and include steam explosion, dilute acid treatment, concentrated acid treatment, ammonia recycle percolation (ARP), treatment with SO_2, treatment with hydrogen peroxide, ammonia fiber explosion (AFEX), and organic solvent treatments. Some of the factors that affect the rate of enzymatic degradation of cellulose include cellulose crystallinity, degree of polymerization, amount of lignin and feedstock particle size. While ARP and

AFEX are among the best in lignin removal, dilute acid pretreatment removes 75% of the hemicellulose.[10,11]

Ionic liquids have been recently shown to be a very promising technique to convert lignocellulosic biomass into amorphous biomass and possibly to fermentable sugars.[12,13] Ionic liquids have low melting points because of the large aromatic or long chain cations and anions with low symmetry and delocalized charge. Most ionic liquids have negligible vapor pressure and have high thermal stability and conductivity. The thermal stability and non-volatility is important from a biofuels point of view because it allows the pretreatment of biomass at ambient pressure. Pressurized equipment cost could be a significant part of the capital cost of a pretreatment plant. Ionic liquids are thus able to dissolve lignocelluloses at ambient pressures in high concentrations and solubilize significant amounts of lignin. These are qualities which are also promising in terms of separating and using lignin as a by-product.

Agricultural residues are composed of 30–40% cellulose, 25–35% hemicellulose, and 10–25% lignin with the remaining percentage in protein, simple sugars, and minerals. However the composition of lignocellulosic biomass is highly variable depending on the geographic area or season. While effective pretreatment is an important step in biomass deconstruction process, pretreatment and subsequent enzymatic hydrolysis are interconnected steps that both play an important role in reducing pretreatment severity and total enzyme loading for economically feasible yields of desired end-products. The most economical biomass conversion technologies will be the one which is insensitive to fluctuations in feedstock properties and processing conditions.

6.4 Cellulases

Cellulose degradation is catalyzed by enzymes called cellulases, which hydrolyze the β(1–4) linkages. Cellulases usually have an alphanumeric nomenclature and are named after the organism that produces them, followed by the family number associated with the catalytic domain, followed by a capital letter that is assigned based on the order in which enzyme family members were discovered in the host organism, alphabetically starting with A.[14] For example, *T. reesei* Cel7A was discovered from *Tricoderma reesei*, is a CBHI with a catalytic domain from Family 7 and was the first cellobiohydrolase of family 7 isolated from *T. reesei*. The enzyme system for the conversion of cellulose to glucose comprises endo-1,4-β-glucanase (EC 3.2.1.4), exo-1,4-β-glucanase (EC 3.2.1.91) and β-glucosidase (EC 3.2.1.21). Cellulolytic enzymes with β-glucosidase act sequentially and cooperatively to degrade crystalline cellulose to glucose. Endoglucanases act in a random fashion on the more amorphous regions of the cellulose fiber whereas exoglucanase or cellobiohydrolases (CBHs) remove cellobiose β(1–4) glucose dimer units from the reducing and non-reducing ends of cellulose chains. Synergism between these two enzymes is attributed to the *endo-exo* form of cooperativity and has been studied extensively between cellulases in the degradation of cellulose in *Trichoderma reesei*, the developmental archetype cellulase system.[15] β-Glucosidase hydrolyzes cellobiose and in some cases cellooligosaccharides to glucose and are generally responsible for the regulation of the whole cellulolytic process. β-glucosidase not only produces glucose from cellobiose but also reduces cellobiose inhibition and allows the cellulolytic enzymes to produce monomeric sugars more efficiently. However, most β-glucosidases are subject to glucose inhibition. The kinetics of the enzymatic hydrolysis of cellulose including adsorption, inactivation and inhibition of enzymes have been studied extensively.[16] Cellulases are very diverse in their structures, mechanisms, and sequences. There are 14 cellulase families listed on the CAZy database (http://afmb.cnrsmrs.fr/CAZY/fam/acc_GH.html) though several of these families (10, 26, 51, 74) mostly contain other types of GHs that are not cellulases. CBHs largely derive from GH families 6,7,9 and 48 although a few are also found in GH family 5. Endoglucanases are found in families 5-9, 12, 44, 45, 48, 51, 61 and 74. β-glucosidases are found in GH families 1 and 3. Cellulases generally show significant sequence homology with some or all of the other family members and utilize the same catalytic mechanism, even if they have different substrate specificities.

6.4.1 Mechanism of Cellulases

Cellulose hydrolysis occurs when cellulases hydrolyze the glycosidic bonds connecting the β-D-glycosyl residues of cellulose. Catalysis involves two amino acids at the active site of the enzyme performing one or two nucleophilic substitution reactions at the anomeric center. The mechanisms of enzymatic catalysis are shown in Figure 6.2. The main difference between the two mechanisms lies in the retention or inversion of the anomeric configuration.

Inversion is a simple single displacement reaction involving a well positioned active site base and acid residues. The catalytic acid residue donates a proton to the leaving group since glycoside hydroxyls have a high pKa and thus are poor leaving groups. The catalytic base is required to deprotonate a water molecule for nucleophilic substitution at the anomeric centre. The acidic and basic residues are usually situated 7–13 Å apart in order to accommodate the nucleophilic water near the pyranoside ring. The roles of acid and base are typically played by either glutamate or aspartate residues. The retention mechanism is a double displacement essentially as outlined by Koshland.[17] In the retention mechanism, a covalent glycosyl-enzyme intermediate is initially formed, that is then hydrolyzed, via oxacarbenium ion-like transition states. This requires the presence of essential amino acid residues, an enzymatic nucleophile and a catalytic acid/base which first serves as a classical Brønsted acid, protonating the leaving group to assist departure and then functions as a base and deprotonates the incoming water nucleophile for the second step. The nucleophile and acid/base are usually always found 5–6 Å apart. Since these classes of proteins have a similar catalytic mechanism, the amino acid residues involved in catalysis are conserved within each family.

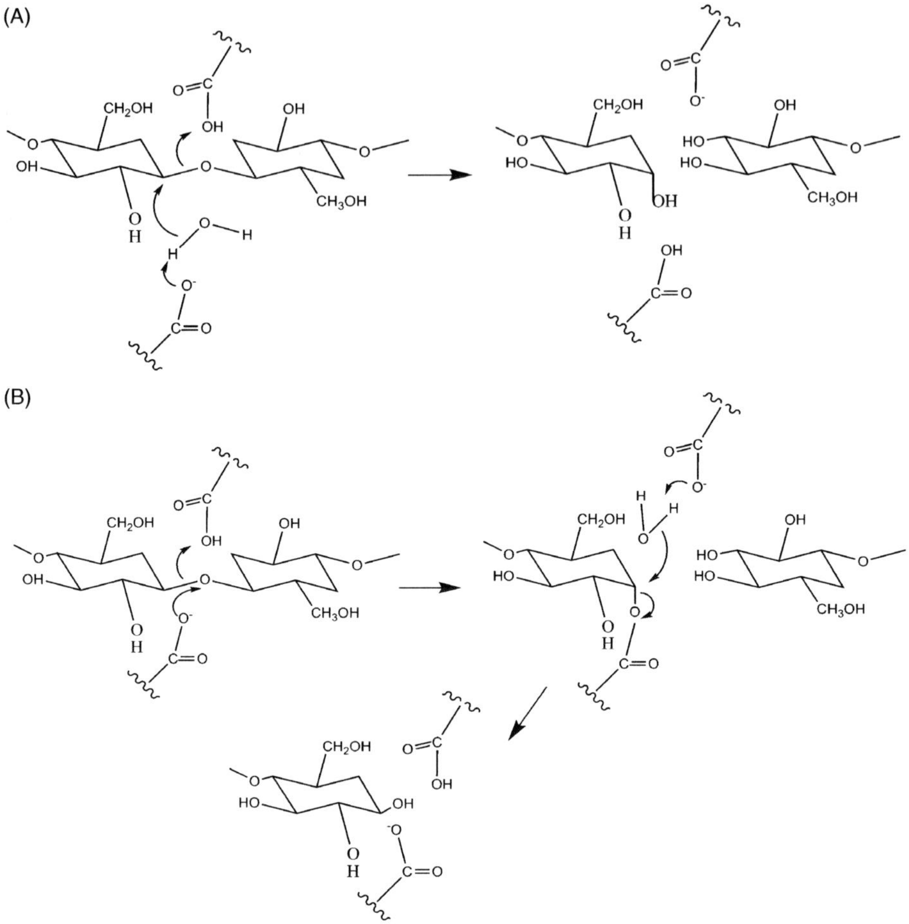

(A)

(B)

Figure 6.2 Catalytic mechanism of Cellulases (A) Inverting mechanism (B) retaining mechanism.

6.4.2 Cellulase Architecture

Most cellulolytic enzymes are modular and contain at least two distinct domains, a catalytic domain and a cellulose binding module (CBM). The catalytic domain hydrolyses the cellulose chain while the CBM is thought to increase the adsorption of cellulolytic enzymes onto insoluble cellulose. Certain aromatic residues of CBM intercalate between the cellulose sheets and helps reduce particle size and increase specific surface area of the substrate. When highly crystalline ramie fibers was treated with a cellulose binding domain and observed under a microscope, small particles released from insoluble cellulose were detected along with an increase in the roughness of crystalline fibers.[18] Most of the early work on cellulolytic enzymes was focused on cellobiohydrolase I (CBH I) from *T. reesei* because CBH I comprises upto 60% of the protein mass of the cellulolytic system of *Trichoderma reesei.*[19] Deletion of the CBH I gene was shown to reduce cellulase activity on crystalline cellulose by 70%.[20] In addition, limited proteolysis studies completed at the same time showed that cellobiohydrolase I was a modular enzyme containing two functional domains.[20] Functional studies showed that the 55 kDa domain contained the catalytic site connected to a small 10 kDa CBM through a linker region. Other examples of structural elements that are found in cellulose structures include multiple catalytic domains, fibronectin 3 like modules and the sequence of hydrophobic residues of cohesins and dockerins found in cellulosomes.

6.4.3 Catalytic Domain

A detailed view of the catalytic domains emerged when the high resolution crystal structures of the CBH I and CBH II were published.[20,21] In cellobiohydrolase I, a 40 Å long cylindrical active site tunnel spanning a third of the catalytic domain is formed by the face-to-face stacking of two large antiparallel β sheets that form a β sandwich to accommodate the cellulose chain. Based on the binding of a cellobioside derivative, this tunnel was proposed to contain the seven glycosyl binding site. The catalytic core of cellobiohydrolase II is made up of α-β fold containing a central β barrel containing seven parallel strands containing four glycosyl binding sites. In both enzymes, two acidic residues, one acting as the proton donor and other as a nucleophile lie above and below the second glycosyl bond of the cellulose chain to release a cellobiose unit of the cellulose chain. The enzyme is then presumed to advance along this chain cleaving of more cellobiose units. The difference between CBH I and CBH II lie in the direction of attack. CBH I bind the non-reducing end of cellulose while CBH II cleaves cellobiose units from the reducing end. The active site architecture and the tunnel also explain why only cellobiose, not any other oligosaccharide, is produced. Endoglucanases however produces a larger variety of hydrolysis products. The crystal structure of endoglucanase I from *T. reesei* confirmed the structural basis of the random attack of the cellulose chain.[22] The absence of the tunnel forming loops in endoglucanase I results in the formation of an open substrate binding cleft. The open active site facilitates the release of many different hydrolysis products. Since the first structures of the fungal cellulases, many new structures with immunoglobin like jelly roll fold, β-sandwiches, (β/α)$_8$ barrels *etc* from diverse sources including bacteria and archaea have been solved.

6.5　Carbohydrate-binding Modules

The inaccessibility of the cellulase active site and the inefficient attachment to the crystalline substrate reduces the efficiency of glycosyl hydrolases. In order to hydrolyze insoluble substrates, these enzymes have evolved to contain catalytic modules that are appended to one or more non-catalytic CBMs. The CBMs play an important role in the ability of these enzymes to degrade crystalline substrates.[23,24] CBMs have been traditionally divided into families based on amino acid sequence similarity. There are currently 45 defined families and these show wide differences in substrate specificity (see http://afmb.cnrs-mrs.fr/CAZY/CBM.html). Fungal cellulase CBMs are mainly found in family 1 while bacterial ones are found in family 2 and 3. CBMs have also been characterized in terms of substrate specificity based on whether they recognize crystalline or non-crystalline cellulose, chitin, β-1,3-glucans and β-(1,3)-(1,4) mixed linkage glucans, xylan, mannan, galactan, and starch. Other CBMs bind to a variety of cell-surface glycans.[23,25–29] Based on the known tertiary structures, CBMs have also been classified into seven fold families with the β-sandwich motif the most common fold. Based on structural and functional similarities, CBMs can be grouped into three types that are discussed below. The reader is also referred to a comprehensive review by Boraston *et al.* from this section was adapted.[24]

6.5.1 Type A Surface Binding CBMs

The Type A proteins binds to highly crystalline cellulose and includes members of CBM families 1, 2, 3, 5, and 10. Aromatic amino acid residues are thought to play a key role in the binding to crystalline substrates.[30] The planar architecture of the binding sites is common to this class of proteins and are complementary to the flat surfaces of the three adjacent glucose residues in cellulose, or chitin crystals.[29,31] While the exact location of the binding site is unknown, the protein is believed to bind to the hydrophobic regions of cellulose. Hydrogen bonding interactions probably do not play a role in substrate binding for these CBMs.

6.5.2 Type B Polysaccharide-chain-binding CBMs

The Type B CBMs include CBMs from 2, 4, 6, 11, 15, 17, 22, 27, 28, 29, 35 and 36[24] and binds to individual glycan chains. Similar to Type A CBMs, aromatic residues and the orientation of these residues play a key role in substrate binding and specificity. The binding site cleft are able to accommodate the individual sugar units of the polymeric substrate.[32] Unlike Type A CBMs, hydrogen bonding interactions play an important role in determining the strength and specificity of glycan chains to Type B CBMs.[33,34]

6.5.3 Type C Small-sugar-binding CBMs

Type C CBMs do not possess the binding site cleft found in Type B carbohydrate binding modules. They bind best to mono-, di-, or tri saccharides and are usually found in xylanases. The type C CBMs includes family 9, 13, 14, 18, 32, 40, and 42. Similar to Type B CBMs, hydrogen bonding interactions between the substrate and ligand play a critical role, and makes the distinction between Type B and Type C CBMs subtle.[35]

6.6   CBM Functions

While CBMs primarily function as accessory modules in polysaccharide degradation they are also involved in complex glucan recognition during glycan synthesis as well as in protein toxin delivery and host colonization by bacterial pathogens.[24] The presence of CBMs in glycoside hydrolases are proposed to be a key factor in their ability to efficiently breakdown insoluble polysaccharides. The three general roles that CBMs have during polysaccharide degradation are proximity effect, a targeting effect and disruptive effect.[24]

6.6.1   The Proximity Effect

CBMs are accessory proteins that bind carbohydrate. In the context of cellulases, the presence of CBM means a tighter association of the enzyme with the polysaccharide thus keeping entire enzyme and the substrate in proximity. When van Tilbeurgh *et al.* treated cellobiohydrolase I from *H. jecorina* with papain to remove the family 1 CBM from the catalytic module[36] they observed that the activity on Avicel was abolished and adsorption of the truncated CBH I to Avicel was correspondingly decreased relative to the full-length enzyme. The reduction of activity on insoluble substrate by the removal of the CBM has been shown in numerous systems including the cellulase family 9 enzymes from *T. fusca*[37] and noncellulolytic enzymes such as amylases, xylanases, and chitinases. The activity of cellulases on soluble substrates are not influenced by CBMs because in such reactions, the reactant molecules are typically well dispersed with the catalytic modules and can make frequent contacts with the substrate. In the presence of insoluble polysaccharides which are not well dispersed, the CBMs probably aid in bringing the enzyme close to the substrate thereby increasing the concentration of the enzyme on the substrate. This proximity effect is also seen in cellulosomes where the glycoside hydrolases are not free in solution but bound to a scaffoldin molecule through protein–protein interactions. The family 3 CBM binds crystalline cellulose and mediates the interaction between the scaffoldin molecule, any adhered enzymes, and cellulose, thereby targeting the cellulosome to the substrate.

6.6.2   The Targeting Effect

In addition to the proximity effect, CBMs are carbohydrate specific and more often than not the specificity of the CBM matches the specificity of the catalytic module.[24] Thus, CBMs also display specificity for polysaccharide substructures. The cellulose-binding type A CBMs are highly specific for crystalline cellulose. In contrast, the cellulose-binding type B CBMs display no affinity for crystalline cellulose but do bind to soluble derivatized cellulose, cellooligosaccharides, and noncrystalline preparations of cellulose. Thus, CBMs are able to discriminate between crystalline and noncrystalline cellulose. When a family 2 CBM was replaced with a family 4 CBM in a crystalline cellulose-specific enzyme, the chimeric enzyme showed a preference for noncrystalline cellulose demonstrating the influence that this specificity can have on enzyme function.[38]

Din *et al.* proposed that there was synergism between the CBM and the catalytic module.[39] The CBM may bind and infiltrate the crystalline cellulose, disrupting and causing release of any particulate, noncovalently-attached cellulose. Together, the CBM and catalytic modules act synergistically and exposing areas of noncrystalline cellulose for hydrolysis. The molecular mechanism of this phenomenon though remains unknown.

6.6.3 Multiple CBMs

Glycoside hydrolases often contain more than one CBM. These CBMs might be from same or multiple families and are not necessarily located next to each other. Previous studies indicate that CBMs in tandem often show increased association constants relative to the individual CBMs because they simultaneously interact with sugar-binding sites that are proximal in 3-D space.[40–42] However this is not always the case. The frequency of multiple CBMs in enzymes has also been observed to correlate with the growth temperatures of the source organisms.[42] This seems to suggest that the presence of multiple CBMs is a way of compensating for the loss of affinity incurred at higher temperatures.[42]

6.7 Enzyme Optimization and Engineering

To produce biofuels economically, the cost of enzymes need to be cheaper and enzyme catalytic efficiency enhanced. The main focus of protein engineering is to increase the cellulose hydrolysis efficiency by improving specific activity, thermostability, pH stability and for maximized synergy of enzymes that are part of a cocktail.

To increase the stability of enzymes, cellulases have been modeled by the SCHEMA energy function[43] using the SCHEMA structure-guided recombination method. In this study, 15 different and diverse thermostable cellobiohydrolase chimeras with upto 7 °C higher melting temperature than the most thermostable parent were found by screening a small library of 73 variants. A high-throughput selection method based on chemical complementation was recently developed by Peralta *et al.* to improve endoglucanase activity. They designing an oligosaccharide surrogate by inserting a cellotetraose between a methotrexate and a dexamethasone.[44] Thus the hydrolysis activity of the endoglucanase was linked to the survival of a URA3-FOA counter-selection yeast strain. This high throughput method yielded two variants with 3.7 and 5.7-fold improved catalytic efficiency from a library size of 10.[8] The selection method used was based on the cleavage of a soluble substrate methotrexate–cellotetraose–dexamethasone. While the method is high throughput, the usage of soluble substrate is a drawback of this technique since the more relevant goal is to engineer cellulase activity toward real crystalline biomass.

Screening against a real substrate requires the development of a high throughput method that incorporates biomass milling and aliquoting. Chundawat *et al.* have developed a high-throughput 96-well microplate technique using solid AFEX pre-treated biomass that was ground and pipetted using ultracentrifugal milling and a robotic multipipetting workstation.[45] While most efforts have focused on engineering individual enzymes, the degradation of biomass in nature is accomplished through the synergistic action of a collection of enzymes. The key to engineering better cellulases would be the ability to engineer enhanced synergy between cellulases and perhaps with hemicellulases and lignin degrading enzymes.

6.8   Cellulosomes

Cellulosomes (Figure 6.3) are extracellular, multi-modular protein complexes produced by a wide range of anaerobic cellulolytic organisms. The main component of the cellulosome is the scaffoldin, a non-catalytic protein sub-unit containing repetitive functional modules called cohesins. The cohesins are involved in specific interaction with protein domains, called dockerins. The cohesin helps incorporate different enzymes and other essential components. The enzymes contain a dockerin domain that binds to the cohesin module and imparts stability to the complex. Scaffoldins sometimes also contain CBMs. Multiple scaffoldins could also be present in the complex. The main distinguishing feature between cellulosomes and cellulases are the presence of the scaffoldin and the dockerin containing enzymes. However recent genome sequencing reports reveal the presence of putative cohesin- and

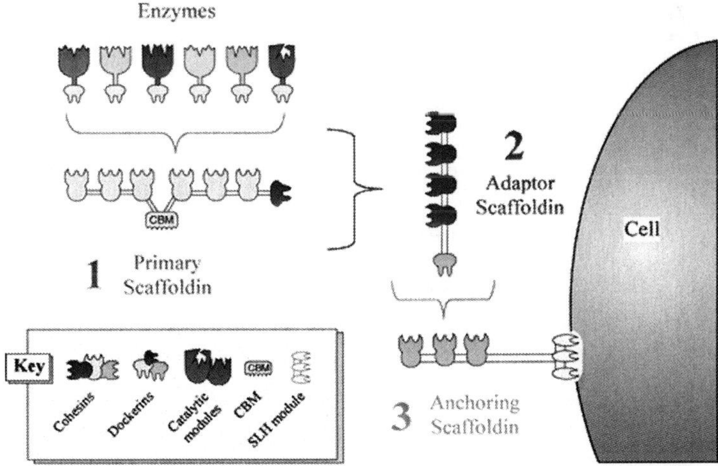

Figure 6.3   A schematic representation of the supramolecular architecture and disposition on the bacterial cell surface of *A. cellulolyticus* cellulosome system. (Reproduced with permission from Annual Review of Microbiology Vol. 58: 521–554).

dockerin-containing proteins in bacteria, archaea, and in primitive eukaryotes.[46] The newly identified modules are however distinct from the classical cellulosome model with most of the parent proteins not predicted to be glycoside hydrolases.[46] The ability to combine more than one type of glycoside hydrolase module in tandem on the cellulosome substructure may allow a synergistic attack on the cellulose or hemicellulose of plant cell walls. For example, the presence of endo- and exoglycoside hydrolases would efficiently degrade a polysaccharide in a concerted manner. Many cellulosomes have varying hydrolytic components, including xylanases, mannanases, lichenases, pectate lyases, chitinases, and cellulases. These glycoside hydrolases may also contain CBMs for recognition of their specific substrates. The architecture of the cellulosomes and the roles of the different components are introduced in the following sections. The reader is also referred to an excellent review by Bayer *et al.* that introduces cellulosomes in great detail[47] and serves the basis for the following introduction.

6.8.1   Non-Catalytic Subunit: Scaffoldin

The first scaffoldin from *Clostridium cellulovorans*[48] was sequenced in 1992. The molecular weight of the cellulosome complex are several megadaltons[49] and are typically made up of two types of subunits. Scaffoldins position and organize the enzymatic subunits and help attach the cellulosome to plant cell wall polysaccharides.[49] The scaffoldins contain multiple copies of *cohesins*, which interact with *dockerin* domains of the downstream proteins through S-layer homology (SLH) modules that mediate attachments to the cell surface. Bacterial cellulosomal systems like *C. cellulovorans* contain a single scaffoldin while *C. thermocellum* contains multiple scaffoldins.

Figure 6.3 depicts the architecture of a cellulosomes in general. The exact architecture varies depending on the identities of the components that make up the cellulosome. The interaction between the cohesin and dockerin are very specific. The dockerin domains comprise about 70 amino acids and contain two 22-amino acid regions separated by a linker region. The first 12 residues of these duplicated sequences are identical to the consensus sequence of the calcium-binding loop in the EF-hand motif.[50] The EF-hand calcium-binding motif is found in calcium binding proteins like calmodulin and troponin C. The cohesins are about 140 amino acids in length and highly conserved in sequence and domain structure. Hundreds of cohesin and dockerin sequences have been found from anaerobic bacteria. Scaffoldins can also be classified into two classes. Class-I scaffoldins contain an internal CBD and a C terminal type II dockerin domain while the Class-II scaffoldins contain an N-terminal CBD without a dockerin domain.

Cohesins are typically made up of 150 amino acids and comprises a nine-stranded β-sandwich motif with a jelly roll topology. *C. thermocellum* cohesins are classified[4] as type-II on the basis of sequence homologies and are different from the type-I on the basis of sequence and binding specificity. The origin of the different classes lies in sequence homologies as well as preferences with regards to the type of dockerin as well as the parent protein. Type-I cohesins include those from anchoring scaffoldin from *B. cellulosolvens* as well as primary scaffoldins from *A. cellulolyticus*.[47] Type-II cohesins are phylogenetically distantly related and include cohesins from *B. cellulosolvens*, *A. cellulolyticus* and *C. thermocellum*. Type-III is primarily found in *R. flavefaciens*. Type-I, II, III cohesin domains bind specifically to the type-I, II, III dockerin domains respectively.

6.8.2 The Cohesin-dockerin Interaction

The biochemical analyses of the cellulosome complex from *C. thermocellum* indicated an exceptionally strong interaction.[51,52] Further analyses supported these claims, and the cohesin-dockerin interaction rates were found to be among the most potent protein-protein interactions known in nature.[53,54]

6.9 Hemicellulases

Unlike celluloses, hemicelluloses are a heterogeneous group of polysaccharides and highly cross-linked to other plant cell wall components. They are linked to lignin via cinnamate acid ester linkages and to cellulose through hydrogen bonds. The linkages to other hemicelluloses are through covalent and hydrogen bonds. Hemicelluloses are heterogeneous and composed of xyloglucans, xylans, mannans, glucomannans and β-$(1 \rightarrow 3, 1 \rightarrow 4)$-glucans. The diversity of hemicellulases is due to the different amounts that are found in different plant families. For example while herbaceous plants contain mainly xylans, in the form of arabinoxylan and glucuronoxylan, softwoods contain glucomannan, galactomannan and galacto(gluco)mannan. As might be expected, such complex structures require a wide variety of enzymes for hydrolysis.

Many hemicellulases are modular enzymes containing catalytic domains, carbohydrate binding domains and dockerin domains that mediate binding of catalytic domains to microbial cell surfaces or cellulosomes through cohesin-dockerin interaction. The catalytic domains are made up of glucosyl hydrolases or carbohydrate esterases.

6.9.1 Hemicellulose Types and Specificity

Hemicellulases can be divided in two major categories: depolymerizing and accessory enzymes and the main types are shown in Table 6.1B, based on information taken from CAZy and BRENDA database.[55,56]

6.9.2 Depolymerization Enzymes

These enzymes act on the hemicellulose sugar chain and can either be endo-acting, like endoxylanases which cleave the sugar chain in the middle or

Table 6.1 Enzymes involved in A. cellulose and B. hemicellulose degradation

Enzymes	Family	Substrate	Products
3.2.1.4 Endo-1,4-β-glucanse 1,4-β-D-Glucan-4-glucanohydrolase	5,6,7,8,10,12,44,45,48,51,61,74	Cello-oligomers	Cello-oligomers
3.2.1.6 Endo-1,3(4)-beta glucanase	16	Cello-oligomers	Cello-oligomers
3.2.1.21 Beta-glucosidase β-D-Glucoside glucohydrolase	1,3,9	Cellobiose	Glucose
3.2.1.91 Cellulose 1,4-cellobiosidase 1,4-β-D-Glucan cellobiohydrolase	5,6,7,9,10,48	Cellobiose	Glucose
3.2.1.37 Exo-β-1,4-xylosidase	3,30,39,43,51,52,54	Xylan	Xylan oligomers
3.2.1.8 Endo-β-1,4-xylosidase	5,8,10,11,16,43,62	Xylan	Xylan oligomers
3.2.1.32 Endo-β-1,3-xylosidase	10,26	β-(1,3)-linked xylan	Xylose
3.1.1.72 Acetylxylan esterase		β-(1,3)-linked xylan O-Acetylated xylan	Xylose Acetic acid
3.2.1.136 Oligosaccharide reducing end xylanase	5	Glucurono-feraxan	Xylan
3.2.1.156 Glucuronoarabinoxylan Endo-1,4-β-1,4-xylanase	8	Xylo-oligomers	Xylose
3.2.1.145 Galactan 1,3-β-galactosidase	43	Arabinogalactan	Galactose

3.2.1.99 Endo-α-1,5-arabinase	43	Linear Arabinan	Arabino-oligomers
3.2.1.55 α–L-arabinofuranosidase	3,10.43,51,54,62	Arabinan, arabinoga-lactan, arabinoxylan	Arabinose
3.2.1.139 α–Glucuronidase	4,67	(4-O-methyl)-glucoronic acid	Glucuronic acid, 4-O-methyl glucoronic acid
3.2.1.22 α–Galatosidase	4,27,36,57	α-Galactopyranose $(1 \rightarrow 6)$ mannooligomers	
3.2.1.89 Endo-galactanase	53	Arabinogalactan Endo-1,4-β-galactosidase	Galacto-oligomers
3.2.1.25 Exo-β-1,4-mannosidase	1,2,5	β-mannosidase	Mannose
3.1.1.6 Acetyl esterase		O-acetylated xylan/ xylo-oligomers	Acetic acid
3.2.1.73 Ferulic acid esterase		Ferrulated xylan	Ferulic acid
3.2.1.78 Endo-β-1,4-mannanase	5,26,4	Mannan, glucomamman galactomannan	Manno-oligomers

exo-acting which work from the chain ends. Endoxylanases are classified both in 3.2.1.8 (β-1$\rightarrow$4) and 3.2.1.32 (β-1$\rightarrow$4) or (β-1$\rightarrow$3) hydrolysis. The exoxylanases are generally found in 3.2.1.37, 3.2.1.72 and 3.2.1.156. As in the case of cellulases, there are enzymes that are both endo and exo acting and the products generated are further broken down by β-glucosidases, β-xylosidases and β-mannosidases. Endomannanases catalyze the hydrolysis of β-D-1,4 mannopyranosyl linkages within mannan as well as glucomannans and galactomannans, the hydrolysis products of which are mannose, mannobiose, mannotrose and so on. Further hydrolysis of mannose requires enzymes like β-mannosidase (3.2.1.25) and β-glucosidase (3.2.1.21).

6.9.3   Accessory Enzymes

These enzymes can be divided based on whether they attack glycosidic linkages or ester-linkages. They include α-glucuronidase, α-L-arabinofuranosidase and α-glucuronidase that remove glucosidic side chains from xylans, and acetyl xylan esterase and feruloyl esterases. α-glucuronidase (3.2.1.139) catalyzes the release of glucuronic acid from xylans. α-L-arabinofuranosidase cleaves arabinose side chains from the xylan chain while α-glucuronidase act on α-galactosyl side chains attached to the mannose backbone. The esterases remove the acetyl and hydroxycinnamic acid substituent's bound to hemicellulose. The acetyl xylan esterases deacytylate xylan and glucogalactomannan while ferulic acid esterases cleave off ferulic acid from arabinose side chain of xylan backbone. Some of these enzymes are very specific for particular oligomers while others act on a variety of substrates.

6.10   Thermophilic Cellulolytic and Hemicellulolytic Enzymes

The variability of the biomass composition requires a robust enzymatic fermentation process that is insensitive to biomass composition and also the pretreatment method. Microorganisms that grow at extreme temperatures could potentially play a key role in biomass conversion processes because of the ability of these organisms to tolerate extreme conditions. Extremophiles which are extreme thermophiles, are microorganisms that grow optimally at 70 °C and higher. Extremophiles survive and thrive in the harshest environments on earth previously thought to not support any forms of life. Recent progress in sequencing technology and the availability of new genome sequences suggests the availability of many putative glycoside hydrolases in such organisms.[57] Till date, cellulosomes have not been found in hyperthermophilic bacteria.

  C. saccharolyticus, a cellulose degrading microorganism with an optimum growth temperature of 70 °C contain a putative bifunctional cellulase CelB (Csac1078), composed of an N-terminal endoglucanase catalytic domain of family GH10, a triplet of CBM3, and a C-terminal exocellulase catalytic domain of family GH5.[57,58] CelA of the same organism (Csac1076) contains a GH9 endocellulase domain, a triplet of CBM3s, and a GH48 exocellulase. The presence of CelA and CelB in *C. saccharolyticus* is thought to facilitate the organism degradation of cellulose.[57]

Extremophilic organisms also contain cellulases that do not contain CBMs. Though *T. maritima* which has a temperature optimum of 80 °C does not grow on cellulose, its genome encodes several β-1,4-glucanases such as Cel5A, Cel5B, Cel12A and Cel12B. All of these enzymes lack CBMs and multiple catalytic domains. *T. maritima* grows on xylan, and its genome encodes two xylanases, XynA and XynB, that contain CBMs.[59,60] The hyperthermophilic archaeon *P. furiosus*, which grows optimally near 100 °C but does not grow on cellulose or xylan, also contains one endo-1,4-glucanase (EglA), but no CBM.[61]

Bacteria contain many thermostable hemicellulases too. As described earlier xylanases from GH family 10 are found in *T. maritima* with XynB having an optimal temperatures of around 90 °C with a broad range of pH. Thermophilic xylanases are also common in fungi. Thermostable β-D-xylosidases have been isolated from both fungi and bacteria.[57]

6.11 Biochemical Conversion of Sugars to Biofuels

The efficient conversion of biomass to biofuels requires a conversion of all sugars produced from the biomass. The ideal organism should be able to convert all sugars and be resistant to any kinds of chemical by-products

Table 6.2 List of thermophilic archae and bacteria and optimal conditions. (Adapted from Blumer-Schuette *et al.*[57])

Organism	Location	T_{opt} (°C)	Source
P. abyssi	North Fiji basin	96	Sea water
P. furiosus	Vulcano Island, Italy	100	Marine sediments
P. horikoshii	Okinawa Trough	98	Sea water
T. kodakaraensis	Kodakara Island, Japan	85	Marine sediments
S. solfataricus	Pisciarelli Solfatara, Italy	85	Solfataric hot spring
T. maritima	Vulcano Island, Italy	80	Marine sediments
T. neapolitana	Lucrino, Italy	80	Marine sediments
T. lettingae	Netherlands	65	Sulfate-reducing bioreactor
T. naphthophila	Niigata, Japan	80	Kubiki oil reservoir
Thermotoga sp. RQ2	Ribeira Quente, the Azores	76–82	Marine sediments
T. petrophila	Niigata, Japan	80	Kubiki oil reservoir
T. elfii	Africa	66	African oil field
Ca. saccharolyticus[e]	Taupo, New Zealand	70	Wood from hot spring
A. thermophilum	Valley of Geysers, Russia	75	Plant residues from hot spring
C. thermocellum	Louisiana, USA	60	Cotton bale

produced during the pre-treatment and biomass hydrolysis process. The two main sugars to biofuels conversion routes are biochemical conversion through microorganisms and thermochemical processes using a chemical catalysis step.

Biochemical conversion usually consists of pretreatment of biomass followed by enzymatic saccharification. Separate hydrolysis and fermentation (SHF) involves hydrolysis of biomass and subsequent fermentation as separate reactions. When both are combined in a single unit, the process is called simultaneous saccharification fermentation (SSF). The rationale of combining saccharification and fermentation in a single unit is to prevent inhibition of the hydrolytic enzymes by the reaction end products. Cellulose hydrolysis is the slow, limiting step in this process and SSF typically takes 3 to 6 days.[62] The biofuel product is typically very dilute and has to be distilled to separate out of the broth. In simultaneous saccharification and co-fermentation reaction (SSCF), the pretreated cellulose is initially saccharified with enzymes and the remaining mixture of sugars and unreacted cellulose transferred to a fermenter in a few days. A fermenting microorganism inoculum is added and the sugars fermented to ethanol concurrently with sugar production from the remaining biomass. Consolidated bioprocessing (CBP) combines cellulase production and simultaneous saccharification of lignocellulose with fermentation of the resulting sugars into a single process step mediated by microorganisms. The one pot CBP process offers the potential of lower capital investment compared to processes involving separate cellulase production. While there are no CBP enabling microorganisms currently reported to being used in industrial processes, the possibility of adapting fungi as CBP organisms remains a promising field of study.

6.12 Summary

With increasing population and global economic development, our energy consumption will continue to increase while the available fossil fuel reserves decrease. To meet this energy demand, fuels from renewable feedstocks are expected to play an increasingly important role in supply of liquid transportation fuels. A major economic challenge towards lignocellulosic biomass production is collection, storage and transport to a biorefinery and the ability to convert the lignocellulosic feedstock to a biofuel, using the existing fuel transportation infrastructure. Current research has been devoted to optimizing pretreatment methods, increasing enzymatic hydrolysis efficiencies and improving microbial hosts for fermentation of the sugars to fuels. While there has been a big increase in our understanding of enzyme structure functional relationships, much more needs to be learned to be able to engineer better enzymes, and perhaps biomass specific enzymes. The next decade will hopefully provide more definitive answers on the scale of use of lignocellulosic derived biofuels along with fossil fuels.

Acknowledgements

This work was part of the DOE Joint BioEnergy Institute (http://www.jbei.org) supported by the U.S. Department of Energy, Office of Science, Office of Biological and Environmental Research, through contract DE-AC02-05CH11231 between Lawrence Berkeley National Laboratory and the U.S. Department of Energy.

References

1. *How dependent are we on foreign oil?*, http://tonto.eia.doe.gov/energy_in_brief/foreign_oil_dependence.cfm.
2. Robert D. Perlack, Lynn L. Wright, Anthony F. Turhollow, Robin L. Graham, Bryce J. Stokes and D. C. Erbach, Oak Ridge National Laboratory, Oak Ridge, TN (DOE/GO-102005-2135), Editon edn., 2005.
3. Z. Zhu, N. Sathitsuksanoh and Y. H. Percival Zhang, *Analyst*, 2009, **134**, 2267–2272.
4. R. Kumar and C. E. Wyman, *Biotechnol Bioeng*, 2009, **102**, 457–467.
5. D. Greer, *Biocycle*, 2005, **46**, 61–65.
6. K. H. Gardner and J. Blackwell, *Biopolymers*, 1974, **13**, 1975–2001.
7. Y. Nishiyama, P. Langan and H. Chanzy, *Journal of the American Chemical Society*, 2002, **124**, 9074–9082.
8. Y. Nishiyama, J. Sugiyama, H. Chanzy and P. Langan, *Journal of the American Chemical Society*, 2003, **125**, 14300–14306.
9. N. K. Vyas, *Curr. Opin. Struct. Biol.*, 1991, **1**, 732–740.
10. L. da Costa Sousa, S. P. S. Chundawat, V. Balan and B. E. Dale, *Curr. Opin. Biotechnol.*, 2009, **20**, 339–347.
11. T. Kim, Y. Lee, C. Sunwoo and J. Kim, *Appl Biochem Biotechnol*, 2006, **133**, 41–57.
12. R. P. Swatloski, S. K. Spear, J. D. Holbrey and R. D. Rogers, *J Am Chem Soc*, 2002, **124**, 4974–4975.
13. C. Li, B. Knierim, C. Manisseri, R. Arora, H. V. Scheller, M. Auer, K. P. Vogel, B. A. Simmons and S. Singh, *Bioresour. Technol.*, 2010, **101**, 4900–4906.
14. D. B. Wilson, *Aerobic Microbial Cellulase Systems*, Blackwell Publishing Ltd., 2009.
15. D. B. Wilson, *Critical Reviews in Biotechnology*, 1992, **12**, 45–63.
16. M. R. Ladisch and G. T. Tsao, *Enzyme Microb. Technol.*, 1986, **8**, 66–69.
17. D. E. Koshland, *Biol. Rev.*, 1953, **28**, 416–436.
18. Neena Din, Neil R. Gilkes, Bahar Tekant, R. C. M. Jr, R. Anthony J. Warren and D. G. Kilburn, *Nature Biotechnology*, 1991, **9**, 1096–1099.
19. P. M. Abuja, M. Schmuck, I. Pilz, P. Tomme, M. Claeyssens and H. Esterbauer, *Eur. Biophys. J.*, 1988, **15**, 339–342.
20. C. Divne, J. Stahlberg, T. Reinikainen, L. Ruohonen, G. Pettersson, J. K. Knowles, T. T. Teeri and T. A. Jones, *Science*, 1994, **265**, 524–528.
21. J. Rouvinen, T. Bergfors, T. Teeri, J. K. Knowles and T. A. Jones, *Science*, 1990, **249**, 380–386.

22. G. J. Kleywegt, J. Y. Zou, C. Divne, G. J. Davies, I. Sinning, J. Stahlberg, T. Reinikainen, M. Srisodsuk, T. T. Teeri and T. A. Jones, *J Mol Biol*, 1997, **272**, 383–397.

23. A. B. Boraston, D. N. Bolam, H. J. Gilbert and G. J. Davies, *Biochem. J.*, 2004, **382**, 769–781.

24. A. B. Boraston, A. Lammerts van Bueren, E. Ficko-Blean, D. W. Abbott and P. K. Johannis, in *Comprehensive Glycoscience*, Elsevier, Oxford, Edition edn., 2007, pp. 661–696.

25. K. Sorimachi, A. J. Jacks, M. F. Le Gal-Coeffet, G. Williamson, D. B. Archer and M. P. Williamson, *J Mol Biol*, 1996, **259**, 970–987.

26. K. Sorimachi, M. F. Le Gal-Coeffet, G. Williamson, D. B. Archer, M. P. Williamson and A. J. Jacks, *Structure*, 1997, **5**, 647–661.

27. B. W. Sigurskjold, B. Svensson, G. Williamson and H. Driguez, *Eur. J. Biochem.*, 1994, **225**, 133–141.

28. M. P. Williamson, M.-F. Le Gal-Coeffet, K. Sorimachi, C. S. M. Furniss, D. B. Archer and G. Williamson, *Biochemistry*, 1997, **36**, 7535–7539.

29. Q. Xu, W. S. Adney, S.-Y. Ding and H. E. Michael, in *Industrial Enzymes*, eds. J. Polaina and A. P. MacCabe, Springer Netherlands, Edition edn., 2007, pp. 35–50.

30. B. W. McLean, M. R. Bray, A. B. Boraston, N. R. Gilkes, C. A. Haynes and D. G. Kilburn, *Protein Eng*, 2000, **13**, 801–809.

31. E. A. Bayer, L. J. W. Shimon, Y. Shoham and R. Lamed, *J. Struct. Biol.*, 1998, **124**, 221–234.

32. P. J. Simpson, H. Xie, D. N. Bolam, H. J. Gilbert and M. P. Williamson, *J. Biol. Chem.*, 2000, **275**, 41137–41142.

33. V. Notenboom, A. B. Boraston, D. G. Kilburn and D. R. Rose, *Biochemistry*, 2001, **40**, 6248–6256.

34. H. Xie, H. J. Gilbert, S. J. Charnock, G. J. Davies, M. P. Williamson, P. J. Simpson, S. Raghothama, C. M. G. A. Fontes, F. M. V. Dias, L. M. A. Ferreira and D. N. Bolam, *Biochemistry*, 2001, **40**, 9167–9176.

35. A. B. Boraston, V. Notenboom, R. A. J. Warren, D. G. Kilburn, D. R. Rose and G. Davies, *J. Mol. Biol.*, 2003, **327**, 659–669.

36. H. Van Tilbeurgh, P. Tomme, M. Claeyssens, R. Bhikhabhai and G. Pettersson, *FEBS Lett.*, 1986, **204**, 223–227.

37. J. Sakon, D. Irwin, D. B. Wilson and P. A. Karplus, *Nat Struct Mol Biol*, 1997, **4**, 810–818.

38. J. B. Coutinho, N. R. Gilkes, D. G. Kilburn, R. A. Warren and R. C. Miller, Jr., *FEMS Microbiol. Lett.*, 1993, **113**, 211–218.

39. N. Din, H. G. Damude, N. R. Gilkes, R. C. Miller, R. A. J. Warren and D. G. P. Kilburn, *Proc. Natl. Acad. Sci. USA*, 1994, **91**, 11383–11387.

40. M. Linder, I. Salovuori, L. Ruohonen and T. T. Teeri, 1996, **271**, 21268–21272.

41. D. N. Bolam, H. Xie, P. White, P. J. Simpson, S. M. Hancock, M. P. Williamson and H. J. Gilbert, *Biochemistry*, 2001, **40**, 2468–2477.

42. A. B. Boraston, B. W. McLean, G. L. Chen, A. R. A. Warren and D. G. Kilburn, *Mol Microbiol.*, 2002, **43**, 187–194.

43. P. Heinzelman, C. D. Snow, I. Wu, C. Nguyen, A. Villalobos, S. Govindarajan, J. Minshull and F. H. Arnold, *Proc. Natl. Acad. Sci. U. S. A.*, 2009, **106**, 5610–5615.
44. P. Peralta-Yahya, B. T. Carter, H. Lin, H. Tao and V. W. Cornish, *J. Am. Chem. Soc.*, 2008, **130**, 17446–17452.
45. S. P. S. Chundawat, V. Balan and B. E. Dale, *Biotechnol. Bioeng.*, 2008, **99**, 1281–1294.
46. A. Peer, S. P. Smith, E. A. Bayer, R. Lamed and I. Borovok, *FEMS Microbiology Letters*, 2009, **291**, 1–16.
47. E. A. Bayer, J.-P. Belaich, Y. Shoham and R. Lamed, *Annu. Rev. Microbiol.*, 2004, **58**, 521–554.
48. O. Shoseyov, M. Takagi, M. A. Goldstein and R. H. Doi, *Proc. Natl. Acad. Sci. U. S. A.*, 1992, **89**, 3483–3487.
49. Q. Xu, E. A. Bayer, M. Goldman, R. Kenig, Y. Shoham and R. Lamed, *J. Bacteriol.*, 2004, **186**, 968–977.
50. S. Chauvaux, P. Beguin, J. P. Aubert, K. M. Bhat, L. A. Gow, T. M. Wood and A. Bairoch, *Biochem J*, 1990, **265**, 261–265.
51. R. Lamed, E. A. Bayer and I. L. Allen, Academic Press, Edition edn., 1988, vol. 33, pp. 1–46.
52. R. Lamed, E. Setter and E. A. Bayer, *J Bacteriol*, 1983, **156**, 828–836.
53. H. P. Fierobe, A. Mechaly, C. Tardif, A. Belaich, R. Lamed, Y. Shoham, J. P. Belaich and E. A. Bayer, *J Biol Chem*, 2001, **276**, 21257–21261.
54. A. Mechaly, H.-P. Fierobe, A. Belaich, J.-P. Belaich, R. Lamed, Y. Shoham and E. A. Bayer, *J. Biol. Chem.*, 2001, **276**, 19678–19678.
55. B. L. Cantarel, P. M. Coutinho, C. Rancurel, T. Bernard, V. Lombard and B. Henrissat, *Nucleic Acids Res*, 2009, **37**, D233–238.
56. I. Schomburg, A. Chang and D. Schomburg, *Nucleic Acids Res*, 2002, **30**, 47–49.
57. S. E. Blumer-Schuette, I. Kataeva, J. Westpheling, M. W. Adams and R. M. Kelly, *Curr Opin Biotechnol.*, 2008, **19**, 210–217.
58. F. A. Rainey, A. M. Donnison, P. H. Janssen, D. Saul, A. Rodrigo, P. L. Bergquist, R. M. Daniel, E. Stackebrandt and H. W. Morgan, *FEMS Microbiol Lett*, 1994, **120**, 263–266.
59. S. B. Conners, E. F. Mongodin, M. R. Johnson, C. I. Montero, K. E. Nelson and R. M. Kelly, *FEMS Microbiol. Rev.*, 2006, **30**, 872–905.
60. Z. Jiang, Y. Zhu, L. Li, X. Yu, I. Kusakabe, M. Kitaoka and K. Hayashi, *J. Biotechnol.*, 2004, **114**, 125–134.
61. M. W. Bauer, L. E. Driskill, W. Callen, M. A. Snead, E. J. Mathur and R. M. Kelly, *J Bacteriol*, 1999, **181**, 284–290.
62. G. Stephanopoulos, *Science*, 2007, **315**, 801–804.

Advances in Gasification for Biofuel Production

CHRISTOPHER R. SHADDIX

Sandia National Laboratories, Livermore, CA 94550, USA

7.1   Introduction

Of all of the conversion processes of biomass into fuels, the gasification process gives the widest breadth of potential products, ranging from gaseous fuels, such as hydrogen or syngas, to liquid fuels such as methanol, ethanol, dimethyl ether (DME), gasoline, or diesel. The gasification process is also one of the most flexible conversion processes in terms of acceptable feed material. Figure 7.1 shows a schematic of the predominant process steps and fuel production routes available when gasifying biomass or biomass-derived feedstocks. The key ingredients of this fuel production route include the gasifier, where one or more oxidizing agents are reacted with the biomass feedstock at elevated temperatures in a fuel-rich environment, the syngas cleanup stage, where harmful contaminants of the gasifier product gas are removed or converted to acceptable chemical species, and the fuel synthesis step, where cleaned, compressed syngas is converted to liquid fuels by reacting over an appropriate catalyst.

The fuel synthesis step can be avoided, along with at least some of the need for syngas compression, by producing gaseous fuels, instead of liquids. Cleaned syngas is an appropriate fuel for operating a gas turbine engine, as has been demonstrated for a number of years in integrated gas turbine combined cycle (IGCC) power plants fuelled by coal or coal-biomass mixtures.[1–3] Syngas can also be used directly to produce electricity in solid oxide fuel cells.[4] As an

RSC Energy and Environment Series No. 4
Chemical and Biochemical Catalysis for Next Generation Biofuels
Edited by Blake Simmons
© Royal Society of Chemistry 2011
Published by the Royal Society of Chemistry, www.rsc.org

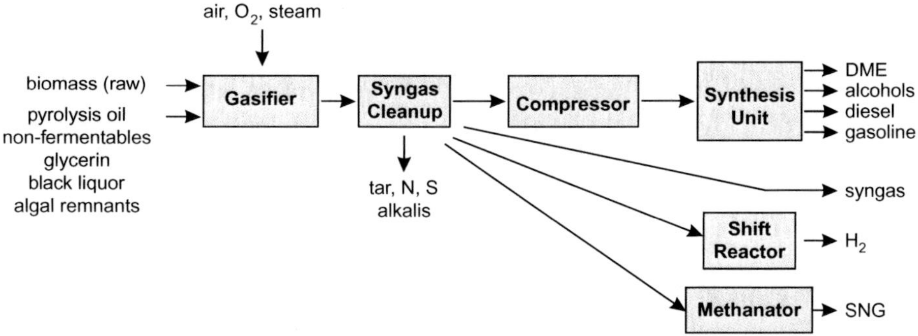

Figure 7.1 Primary process steps and potential fuel products arising from gasification of biomass feedstocks.

alternative, syngas, primarily composed of carbon monoxide and hydrogen, can be converted to high-purity hydrogen by reacting it with steam over one or more shift catalysts and then separating the hydrogen from carbon dioxide.[5–7] Hydrogen can be used to fuel vehicles via internal combustion engines or fuel cells, or can be used to generate electricity through fuel cells or gas turbines.[7–13] Another gaseous fuel possibility is to generate substitute natural gas (SNG) from syngas by passing it over a suitable nickel oxide methanation catalyst.[14] The methanation reactions typically involve 3 or 4 moles of hydrogen reacting with 1 mole of CO or CO_2 (yielding methane and water), so the methanation catalyst must also shift the syngas to the appropriate concentrations of CO, CO_2, and H_2. Coal gasification plants in North Dakota and in South Africa have been generating SNG in this manner for many years.

The generation of liquid fuels from syngas requires that syngas within a certain range of H_2/CO ratio be passed over a catalyst operating at a particular temperature and pressure. As such, a shift reactor is often required to tailor the H_2/CO ratio for generation of the particular liquid fuel that is desired (assuming the raw syngas has a lower H_2/CO ratio than desired). The range of liquid fuels that can be generated from syngas extends from oxygenates such as DME and alcohols to hydrocarbon mixtures such as diesel and gasoline. The simplest synthesis route is undoubtedly to methanol, via reaction over a $Cu/ZnO/Al_2O_3$ catalyst at 35–55 bar and 200–300 °C.[15,16] The ideal H_2/CO ratio for generating methanol is ∼2, as suggested by the overall stoichiometry of the simplified reaction:

$$CO + 2H_2 \rightarrow CH_3OH$$

Methanol is suitable as an automotive fuel in internal combustion engines or molten carbonate fuel cells, but its corrosiveness and affinity for water pose additional costs associated with its storage and handling. Methanol itself can act as a base material for generation of a variety of fuels and chemical feedstocks.[17,18] For application to liquid transportation, the primary products from methanol include olefins (a component of gasoline), via the commercial methanol-to-olefins (MTO) process,[17,18] DME, via catalytic dehydration of methanol,[19]

$$2CH_3OH \rightarrow CH_3OCH_3 + H_2O$$

and even gasoline, via the commercial methanol-to-gasoline (MTG) process.[17,18]

Direct synthesis of DME from syngas is optimized for a H_2/CO ratio of one, according to the reaction:[20]

$$3CO + 3H_2 \rightarrow CH_3OCH_3 + CO_2$$

As suggested by the reaction stoichiometry, this reaction is more difficult to achieve than methanol synthesis and also results in lost carbon (manifest as a CO_2 by-product). DME synthesis reactors are typically operated at 50 bar and 250–260 °C.[20–22] DME is an attractive fuel because its properties are similar to those of liquefied petroleum gas (LPG), and therefore it can be distributed and stored using familiar LPG technology. DME has favorable ignition characteristics (its cetane number is approximately 60) and is therefore appropriate for use in diesel engines, once one makes suitable modifications to the fuel injection system.[23,24] In fact, soot emissions are greatly reduced and NOx emissions can also usually be reduced when operating a diesel engine on DME. It also is an appropriate fuel for use in gas turbines for power generation.[25]

This chapter summarizes the current status and prospective areas for improvements in the biomass gasification process itself. As discussed above and illustrated in Figure 7.1, the gasification process is a key component of a wide range of production routes for both gaseous and liquid biofuels. Improvements in the efficiency, capital effectiveness, and availability of the gasification process will have important impacts on the overall cost effectiveness of these thermo-chemical routes of biofuels production and may therefore play a significant role in the anticipated expansion of biofuels production in the future. The chapter begins with a discussion of the diverse range of biomass feedstocks appropriate for use in gasification processes and then progresses to the core topic: the current state-of-the-art in biomass gasification. Implications of the different gasification approaches on gas cleanup requirements are then discussed, before summarizing the topics that have been covered and forecasting the expected areas of improvements in gasification technology in the future.

7.2   Biomass Feedstocks for Use in Gasifiers

As suggested by the wide range of biomass-derived feedstocks entering the biofuels production process layout in Figure 7.1, many different feedstocks are suitable for use in gasifiers. In fact, because most biomass sources have very similar atomic ratios of carbon, hydrogen, and oxygen and because the gasification process reacts the fuel source down to the smallest possible molecular units (CO and H_2, with some CH_4 for certain gasification processes), the syngas composition leaving the gasifier is itself fairly insensitive to the type of biomass feedstock that is being gasified. This 'feedstock-neutrality' is a significant advantage of gasification-based conversion (as well as other thermochemical processing routes, such as pyrolysis) relative to biochemical conversion, which typically is most effective for starch or cellulose and has difficulty converting hemicellulose, lignin, and protein components of biomass.

The two primary limitations to acceptable feedstocks for gasification are the water and ash contents of the feed. Excessive water in the feedstock lowers the system efficiency because of heat losses associated with first vaporizing and then recondensing the water that is fed with the fuel into the gasifier. Similarly, excessive ash components act as a heat sink in the gasification process and can lead to bed material agglomeration and slag management problems, as will be discussed in more detail in Section 7.3. The feedstock water content can be minimized by applying feedstock drying practices, including the use of low-quality waste heat that is generated at the biofuels plant, particularly in the syngas cooling, compression, and biofuels synthesis steps. The ash content of the feedstock can also be minimized by applying stone removal and density fractionation steps to remove extraneous mineral matter.

Pre-processed biomass streams, including black liquor (a byproduct of pulpmaking), glycerin (a byproduct of biodiesel production), lignin residues from cellulosic ethanol production, pyrolysis oil, and algae or algal oils, are particularly suitable for use in gasification-based biofuels production because they are generally liquids and can be readily injected into a pressurized gasifier. Pressurizing the gasifier can have substantial benefits to the overall process efficiency and economics, as will be discussed below. Unfortunately, for most of these pre-processed biomass streams, at least part of their fluidity results from their water content, which, as mentioned previously, results in a less efficient gasification process.

7.3   Gasification Technologies

Fundamentally, the word 'gasification' refers to the overall conversion of a solid or liquid fuel feedstock into gaseous fuel. For non-charring fuels, however, simple liquid spray evaporation or simple solid fuel thermal devolatilization is generally not considered to be a gasification process. Rather, chemical reactions that convert the solid fuel to gaseous form are an essential component of gasification. Most biomass fuels, in contrast to coal, undergo extensive devolatilization as they are heated to high temperatures (typically losing 70–80% of their mass), which means that the amount of char remaining that needs to be chemically gasified is relatively low. In fact, the extent of biomass devolatilization is a strong function of the heating rate, final temperature, and pressure of the process,[26] so different approaches to biomass gasification can have considerably different biomass char yields within the gasifiers. Because of this strong role of devolatilization in the conversion of the biomass in gasifiers, the pyrolysis and steam reforming reactions of the biomass volatiles play an important role in the overall chemical composition of the gasifier product gas.

A wide variety of gasification technologies have been developed over the years for biomass, owing to the wide variations in feed size, moisture, particle density, etc. for different sources of biomass, as well as wide variations in the desired reactor throughput, based on the magnitude of the available biomass resource. Most of these technologies have involved air-blown gasification, which is relatively simple and low-cost, but which produces a low-energy content and highly diluted 'producer gas' that is inappropriate for subsequent fuel synthesis. To produce a fuel synthesis-quality product gas, or 'syngas', three general approaches for biomass gasification have been developed. One of these approaches uses steam and an external source of heat (typically generated in an adjacent combustion reactor that is fuelled by residual char) to gasify the biomass, in a moderate temperature fluidized bed.[27] A second approach uses a sub-stoichiometric quantity of oxygen, produced via cryogenic air separation or a pressure swing adsorption technique, to gasify the biomass. This approach is usually performed at high temperatures, in an entrained flow process that requires significant communition of the biomass feed source,[28–30] but can also be applied in a fluidized bed, if the oxygen is sufficiently diluted with steam before injection.[29,30] A third approach, originally developed for application to waste treatment that has seen considerable interest recently, involves the use of high-temperature plasma sources.[31] After a discussion of the important operational criteria of gasifiers, these three different approaches to biomass gasification will be described in some detail below, before progressing to a discussion of the syngas cleanup processes, as the required cleanup processes are closely linked to the type of gasification process employed. It should be noted that hydrothermal gasification, which occurs at low temperatures and high pressures in an aqueous-phase reactor, produces a product gas which is predominantly composed of CO_2 and methane and is not appropriate for use for fuel synthesis.[32]

7.3.1 Operational Characteristics of Gasifiers

Important considerations in the operation of gasifiers include the gasifier pressure, temperature, and energy balance. The pressure is an important factor with regards to feeding the biomass stream into the reactor, the required size of the reactor, and the required size of the subsequent gas cleanup equipment. For fuel synthesis applications, the syngas ultimately needs to be compressed (usually to quite high pressures) for catalytic synthesis, so it is beneficial to operate the gasifier at an elevated pressure, because the introduction of the solid fuel at pressure does not require the compression cost that is entailed for the gaseous products of the gasification process. Also, the sorbent-based and catalytic gas cleanup processes can be reduced in size and are less expensive when operating at pressure. The gasifier size can probably be reduced in size when operating at pressure, because the gasification reactions themselves progress more rapidly at higher pressures (because the collision rate of gas reactants with the solid char increases with pressure, for a given temperature). However, the amount of volatiles produced decreases with increasing pressure (and therefore the amount of char remaining to be gasified increases). The competition between these two competing factors has not been extensively investigated for the different gasification processes. Finally, the gasifier pressure plays an important role in the ease with which a given biomass feedstock can be fed into the gasifier, specifically in the extent of communition of the feedstock and appropriate moisture level for consistent feeding. As the gasifier pressure increases, greater size reduction and uniformity of the biomass feed is required, and at high enough pressures the biomass may need to be introduced as a fuel slurry in water.

The gasifier temperature dictates the rate of devolatilization and char gasification reactions, the overall extent of devolatilization, the degree of softening of the inorganic elements in the biomass (particularly for the alkali metals and their associated salts), and the extent of corrosive attack of refractory and metal surfaces. Gasification processes generally operate above or below the temperature range where the primary inorganic components of the fuel transition from hard solids to molten liquids (slag). This transition generally occurs over a temperature range of approximately 100–200 °C for most types of biomass, but the actual beginning and endpoint temperatures of the slag transitions vary substantially, particularly as a function of the alkali and chlorine content of the biomass (most biomass sources have slag transitions falling within the range of 850–1200 °C).[33–35] If operating below this temperature, the biomass devolatilization process is less effective at generating volatiles and the char gasification reactions proceed very slowly. Consequently, long residence times are required to yield acceptable conversion of the char, implying the use of some type of bed-based gasifier. Fixed bed gasifiers are relatively simple to operate and can be operated in either an upflow or downflow configuration. Upflow, or countercurrent, gasifiers do a good job of minimizing energy loss through the produced gas, but suffer from high concentrations of methane and tars produced from pyrolysis of the fuel volatiles. Downflow, concurrent gasifiers have much lower concentrations of methane and tars in the produced gas, but suffer from extensive heat loss through the sensible heat of the produced gas. For both types of fixed bed reactors, the temperatures and stoichiometries can vary significantly throughout the bed, making it difficult to have a consistent product gas composition.[36] The use of moving beds and, especially, a fluidized bed, improves the mixing throughout the bed and enhances the fuel heating and devolatilization processes, but at the cost of increased complexity and flow control requirements. Also, the feedstock needs to be processed to an appropriate, fairly uniform particle size for use in a fluidized bed. As fluidized beds necessarily operate in an upflow, countercurrent geometry, tar levels are quite high in the produced gas.

Slagging gasifiers operate at temperatures above the flow point of the mineral constituents of the fuel, such that the minerals predominately flow out of the gasifier as molten slag (unfortunately, some of the alkali metals vaporize and exit the gasifier as an alkali fume at these process conditions). Under these conditions, the introduced particles usually are gasified in an entrained flow process and therefore they must be processed to be sufficiently small to be carried by the gas flow and to be gasified within the available residence time within the reactor. At these temperatures the tar components formed by the biomass volatiles rapidly undergo pyrolytic reactions that result in production of small gas compounds and soot, and the hydrocarbon species and soot particles rapidly react with steam and CO_2 to yield the desired syngas components, H_2 and CO. Unfortunately, vaporized and liquid alkali constituents actively attack traditional refractory materials at these temperatures, making refractory repair and replacement a regular occurrence in this type of gasifier. It should be noted that most commercial coal gasifiers are oxygen-driven entrained flow gasifiers, with a wealth of operating experience.

A critical operational variable in gasification is the energy balance. For a steady-state process, the energy losses from the gasifier must be balanced by the energy inputs. The biomass introduced to the process must first be heated and undergo devolatilization (a slightly endothermic process) and then have its char at least partly converted through gasification reactions with steam and CO_2. The gasification reactions are strongly endothermic, as shown in Equation 1 and 2.

$$CO_2 + C(s) \rightarrow 2CO \qquad \Delta H = +172 \text{ kJ/mole C} \qquad (1)$$

$$H_2O + C(s) \rightarrow H_2 + CO \qquad \Delta H = +130 \text{ kJ/mole C} \qquad (2)$$

Furthermore, the reactor itself loses heat through its walls and through the sensible enthalpy of the hot gas stream and ash leaving the reactor. There are two methods to provide these heat requirements, which lead to the characterization of the process as either direct gasification or indirect gasification. In direct (or autothermal) gasification, the necessary heat is provided internal to the gasifier by using air or oxygen to the reactor and relying on the localized combustion heat release to drive the gasification process, as shown in Equations 3 and 4 (depending on the temperature and pressure, the char will tend to oxidize more to CO or to CO_2).

$$O_2 + 2C(s) \rightarrow 2CO \qquad \Delta H = -110 \text{ kJ/mole C} \qquad (3)$$

$$O_2 + C(s) \rightarrow CO_2 \qquad \Delta H = -394 \text{ kJ/mole C} \qquad (4)$$

Another option is to provide the necessary heat energy through the transfer of one or more hot mediums (e.g. steam and sand) introduced to the reactor. This process is known as indirect (or allothermal) gasification. An advantage of indirect gasification is the complete absence of diluent gas and combustion products in the produced syngas, as are found in direct gasification, especially for air-blown direct gasification. Plasma gasification is generally applied as a combination of allothermal and autothermal approaches, wherein the electrically generated plasma heat is introduced to the reactor, together with some autothermal oxygen.

7.3.2 Moderate Temperature, Indirect Gasification

The indirect, dual-bed biomass gasification process was derived by analogy to the fluidized catalytic cracking (FCC) process that was developed in the late 1930s in the petroleum refining industry.[27,37] This biomass gasification process has seen the most extensive development by far of the three characteristic approaches to generating biomass syngas. In the FCC process, catalytic material reacts with fuel vapors, resulting in coking of the catalyst. To regenerate the spent catalyst, it is removed from the main reactor and the coke layer is burned off in a regenerator reactor (with an air feed), before mixing with the fresh hydrocarbon feed going into the main reactor. The extra heat of the regenerated catalyst is used to vaporize the fresh feed and maintain the temperature in the main reactor.[37] In the indirect gasification system, biomass feed is partially gasified with steam and a solid heat transfer medium (typically some form of sand) in the main reactor to produce syngas and char. The char and sand is then separated from the syngas and fed into an air-blown combustion unit that burns off the char to heat the sand and produce steam, with the steam and hot sand returned to the main reactor (see Figure 7.2). The details of the operation of the two linked reactors have varied considerably over the nearly 30 years of implementation of this approach by various groups around the world, as detailed in a recent review article by Corella *et al.*[27] In some cases circulating fluidized beds have been used for both the gasification and combustion reactors (as in the Battelle and Battelle/FERCO designs),[38] whereas in other cases bubbling fluidized beds or risers have been used for one or for both reactors. In general, the gasification reactor has been operated at a temperature of 700–800 °C, whereas the combustion reactor has been operated at 850–950 °C. Both reactors are generally operated at pressures just slightly above atmospheric pressure (considered 'low pressure' operation, in comparison to most other gasification technologies). Under these conditions, the conversion of the biomass feed in the gasifier is dominated by devolatilization, and actual heterogeneous reaction of the resultant biomass char with steam is

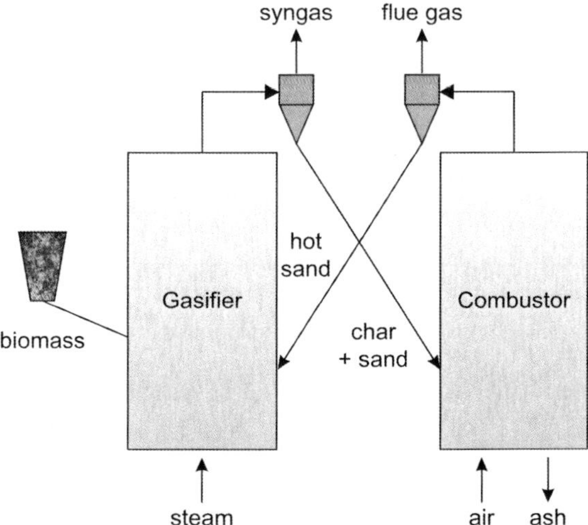

Figure 7.2   Schematic of a basic dual-bed, indirect gasifier.

very slow.[39,40] Tar production also tends to be strong under these gasification conditions, so many different reactor configurations and bed additives have been investigated in attempts to improve char conversion and reduce tar formation.[39–43] The largest indirect gasification plant built to-date has been the Battelle/FERCO gasifier that was installed and briefly operated in Vermont, which had a capacity of nearly 200 tonnes per day (tpd) of dry wood feed.[38]

Table 7.1 lists a typical syngas composition produced through indirect gasification.[27,30,42,44,45] The composition shown here is most reflective of the operation of the Battelle gasifier, as this one appears to have the most thoroughly documented experimental results and has been scaled to the largest size. In fact, in indirect gasification a wide range of H_2/CO ratios, in particular, can be generated, depending on the steam content in the gasifier and the gasifier temperature, pressure, and residence time. This is because of the important role of the water-gas shift reaction in determining the balance of the major species:

$$H_2O + CO \leftrightarrow CO_2 + H_2 \tag{5}$$

At higher temperatures, CO and steam are favored at the expense of CO_2 and H_2, and vice versa at lower temperatures. As is evident in Table 7.1, the indirect gasification approach produces a good quality synthesis gas, at least on a dry basis, if one can effectively convert the abundant hydrocarbons that are produced in the process and either convert or remove the abundant tar. The high moisture level that necessarily accompanies the syngas in this process represents a significant efficiency loss in the gasification system.[27] For fuel synthesis applications, this moisture will be mostly removed through successive compression and cooling stages before fuel synthesis, but most of this low-quality heat from steam condensation cannot be effectively recovered.

In addition to the traditional dual-bed approach to indirect gasification, a couple of approaches have also been developed that aim to simplify heat and mass transfer through the insertion of hot tubes that pass through the gasifier

Table 7.1 Characteristics of biomass gasifier product gas.

Gasifier type	Product gas composition (vol-%, dry)							H_2/CO	LHV MJ/Nm³ [*]	Efficiency
	CH₄	*C₂*	*H₂*	*CO*	*N₂*	*CO₂*	*tar*			
Indirect, dual-bed	14	5	24	40	1	16	0.1[†]	0.6	16	0.70
Direct, O₂-blown bed	13	3	29	22	3	30	0.05	1.3	12	[‡]
Direct, O₂-blown entrained flow	0.1	0	34	37	3	26	0	0.9	10	[‡]
Thermal plasma	0.5	0	37	60	0	2	0	0.6	12	[‡]

[*]units of megajoules per normal cubic meter, with normal conditions defined to be 1 atm pressure and 298 K.
[†]0.1% tar corresponds to ∼30 g/Nm³.
[‡]gasifier efficiency cannot be meaningfully defined for those processes that utilize oxygen and/or plasma heating sources.

bed. In one case, the tubes contain the hot flue gas from char combustion, while in another the tubes act as pulse combustors that consume a portion of the syngas exiting the gasifier.[30] In this latter case, effective gasification of the char is required in the gasifier, so the requisite bed depth is much larger than in traditional indirect gasifiers. Neither of these approaches appears to have any active development activity at this time.

Several approaches to indirect gasification have been recently developed which have focused on better process thermal integration, to reduce the heat losses and attendant efficiency loss associated with transfer of hot solids between two separate reactors. One approach uses concentric fluid beds and another uses neighboring fluid beds with a common wall and internal solids circulation.[27,46] As emphasized by Corella *et al.*,[27] such systems with internal solids circulation tend to suffer from gas leakage between the oxidizing and reducing zones. A third approach is to use a central riser within a fluidized bed. Xu *et al.*[42] compared the performance of gasification in the riser surrounded by char combustion in a fluid bed and the converse, and found improved performance when char combustion occurs in the high velocity riser, as one would expect, based on the much faster char oxidation process. Based on both measurements and process analysis, cold gas efficiencies of 70–75% are attainable with this approach.[39,42] Xu has also been investigating the implementation of a two-stage gasification approach in which the hot sand particles leaving the char combustor are conveyed first to a secondary fluid bed in which the hydrocarbon-laden gas exiting the primary gasifier stage is partially reformed, thereby increasing the cold gas efficiency and lowering the tar yield.[43] A variant of the riser-in-bed approach has been developed at the Energy Research Center of the Netherlands (ECN) and is known as the Milena gasification technology.[14,29] In this approach the biomass is gasified in a central riser that is contained within a bubbling fluid bed combustor that converts the char particles and recycles the sand. Recently, this technology was scaled up to a level of 4 tpd.[14] Finally, Iliuta *et al.*[47] have proposed a multi-compartment integrated gasifier/combustor concept that benefits from increased wall heat transfer associated with larger surface-to-volume ratios of smaller fluidized bed compartments and from periodic cycling from oxidating to reducing environments in each compartment (eliminating the need to physically move the char from one bed to another). While philosophically appealing, the challenges associated with practical implementation of this approach, particularly regarding biomass feeding and segregation of syngas and oxidizing flue gas, appear formidable.

In summary, in comparison to other gasification technologies for producing synthesis gas, the indirect gasification approach is fairly mature and robust. It can be scaled relatively easily to high processing volumes and has no difficulty with biomass feed. However, application of this technology to fuel synthesis has not been demonstrated, and there are strong concerns with respect to its low-pressure operation (requiring syngas compression), its capital-intensive nature, with two interconnected bed reactor systems, and the high concentrations of hydrocarbons and tars in the syngas, requiring secondary upgrading.

7.3.3 Oxygen-blown Direct Gasification

In comparison to indirect gasification, there has been little historical experience in the oxygen-blown direct (autothermal) gasification process, mostly because the high cost of oxygen production has made application of this technology for heat and power production (the traditional end-products of biomass gasification) untenable. However, the current interest in synthetic liquid fuel generation and the promise of future cost reductions in the cost of oxygen generation make this gasification approach of significant interest.

As mentioned previously, there are two basic variants of direct gasification with oxygen. The moderate temperature approach is modeled after traditional air-blown bubbling fluid bed or circulating fluid bed gasifiers and simply uses steam-diluted oxygen to simulate air as the gasifying agent. In this application the steam must be at least partially premixed with the oxygen before injection into the bed, to avoid the creation of local hot-spots and clinker formation from partially molten inorganic material. Because this gasification process takes place in a single reactor vessel and uses oxygen, which is available at pressure, it can be easily operated at pressure and has been demonstrated up to a pressure of ~ 20 bar. Pressurizing the gasifier reduces the costs of subsequent gas cleanup and syngas compression. Production of hydrocarbons and tar is still significant, because of the moderate temperature of the process, so secondary reforming and/or cracking of the product gas is necessary. Table 7.1 shows a typical gas composition exiting this type of gasifier.[30] Note that the autothermal nature of the gasification process results in a significant concentration of CO_2 in the product gas, on account of partial oxidation of the fuel (to release the energy needed to drive the gasification reactions). This is the primary difference in composition of the direct gasification syngas in comparison to syngas generated from indirect gasification. The carbon conversion in these systems approaches 90%.[29]

The second approach for direct gasification with oxygen is known as entrained flow gasification. This type of gasifier has, in fact, seen the widest commercial utilization, though primarily for application to pulverized coal and/or petroleum coke.[48] In an entrained flow gasifier the fuel particles are carried along (*i.e.* entrained) in a high temperature gasifying medium and must be gasified during their short residence time being carried through the reactor. As a consequence, the fuel particles must be small (less than 0.1 mm in size, for coal particles), though not necessarily dry – both dry-feed and liquid-slurry feed systems have been employed for coal particle feeding. High temperatures are required to accelerate the gasification reactions to consume the fuel particles within the available residence time. Consequently, these gasifiers are generally operated as slagging gasifiers, wherein the mineral components of the fuel melt, flow along the walls and bottom of the reactor, and leave the reactor as molten slag. The slag layer also acts to insulate and protect the gasifier refractory wall from corrosion. The entrained flow gasification of biomass, with a significant fraction of its inorganics composed of low-melting point alkali metals, can potentially be conducted at somewhat lower temperatures than have traditionally been applied for coal, 1300–1500 °C, which would help to reduce refractory corrosion. Entrained flow gasifiers are typically operated under pressurized conditions of 20–80 bar.

Entrained flow gasification of biomass has only been demonstrated at a commercial level by co-gasifying biomass with coal in existing coal gasifiers. A small percentage of biomass (5%) has been co-gasified with coal in the slurry-fed Tampa Electric Polk Power Station (220 MW_e) in Florida and a significant fraction (30%) of biomass has been co-gasified with coal in the dry-feed Nuon Buggenum plant (250 MW_e) in the Netherlands. Slurry feeding of biomass is constrained by the hygroscopic and hydrophilic nature of the biomass constituents, which limits the solids loading in slurries.[49]

For applications to co-gasification, the biomass has been reduced to sufficiently small sizes by mixing the biomass together with the coal before sending it through the coal pulverizers. While the biomass is not reduced down to the same size range as the coal, its high volatile content (which results in strong particle shrinkage upon devolatilization) and the relatively high reactivity of biomass chars means that the biomass only needs to be reduced down to a characteristic size of ~ 1 mm. For a dedicated biomass entrained flow gasifier, traditional biomass milling equipment cannot reduce the particle size down to this level without extensive feed recycling and unacceptable costs. This has posed one of the major barriers (the other being the cost of oxygen) to commercial implementation of entrained flow gasification of biomass. However, during the last few years some concepts have arisen to address this shortcoming. One approach is to pyrolyze the biomass as a preprocessing step and then to use the pyrolysis oil that is generated as a liquid carrier for pulverized pyrolysis char. In this way the biomass can be introduced into the gasifier much as heavy oil (or slurried coal) is in the Texaco gasifier design. Alternatively, if there is sufficient value in other use of the pyrolysis-generated char (for example, as a filtration medium or as a soil amendment), one can simply spray the raw bio-oil into the gasifier. The other approach to biomass preprocessing that shows some promise is to torrefy the biomass and then feed the pulverized torrefied biomass (and the volatiles released during torrefaction) into the gasifier.[50]

A typical syngas composition associated with entrained flow gasification of biomass is shown in Table 7.1. As with the other direct gasification syngas, a significant amount of CO_2 is in the product gas. Unlike all of the previously considered (bed-based) gasification techniques, the production of hydrocarbons and tar is negligible in syngas from entrained flow gasification, because of the high temperature of the process, which results in complete reforming of the biomass volatiles into CO and hydrogen. Also, with the high temperatures in entrained flow gasification, carbon conversion can exceed 99%.

All of the direct gasification approaches described in this section rely on the use of oxygen. For a large-scale coal gasifier, the high capital cost associated with cryogenic air separation units (which liquefy air and then distill out the oxygen) can be recovered in a reasonably short time. Smaller, biomass-based gasification systems cannot afford to have such a high capital cost, and the energy efficiency of air separation with cryogenic methods decreases significantly with smaller plant sizes. Consequently, one can use liquid oxygen, carried by truck or rail to the gasification facility (and presumably produced by cryogenic air separation at a major gas handling facility), or use smaller-scale, lower capital approaches to on-site oxygen production. Membrane-based approaches to oxygen production (e.g. the high temperature ion transport membrane approach) have been in development for a number of years, but still seem to be far away from commercial readiness. Therefore, the most logical approach at this time is to use a pressure swing adsorption (PSA) technique for oxygen production. In this approach, compressed air is passed over an adsorption column that preferentially adsorbs nitrogen, especially, and argon, allowing a fairly pure stream of oxygen to pass through and exit the bed. Once the adsorption media is saturated, the airflow is shut off and the trapped gases are vented before initiating another cycle. By using multiple adsorption columns operating in parallel and, occasionally, in series, a continuous flow of high purity oxygen (typ. 95% pure) is produced.[51] Further efficiency improvements have recently been achieved with this technology by venting the trapped gas to a vacuum, in a process known as vacuum PSA (VPSA).

7.3.4 Plasma Gasification

The thermal plasma gasification process has been developed over a number of years for treatment of carbonaceous waste streams.[52–54] Essentially this is the same technology as thermal plasma pyrolysis, except it is conducted in the presence of steam, allowing for rapid steam-reforming reactions of the hot pyrolysis products and gasification reactions of the steam with char. In this sense, the thermal plasma gasification process combines the allothermal characteristic of indirect gasification (except the thermal source is externally supplied electricity) and the high temperature volatiles yield and active char gasification characteristic of oxygen-blown entrained flow reactor gasification. Because of the high temperatures associated with thermal plasmas, there is the possibility of producing a product gas with negligible tar content. The high power densities and high temperatures associated with thermal plasmas also offer the possibility of a large reactor throughput in a small volume. The key tradeoffs for these positive aspects of plasma gasification are the high electricity costs and electrode wear and replacement.

It should be noted that non-equilibrium (or 'cold') plasma processes have seen substantial development in recent years for various solid etching or other surface modification processes. These types of plasmas can be generated at relatively low temperatures and generally have lower electrical power consumption than thermal (equilibrium) plasma sources. However, to keep the plasma in non-equilibrium, it must necessarily operate at low gas density, implying reduced pressure. The large amount of residual char produced under these conditions[55,56] and the additional compression of the product gas required before gas cleanup would appear to make this approach to plasma gasification uneconomical.

While a large amount of research and development has taken place in the development of cost-effective waste treatment using thermal plasma techniques, research on production of syngas from biomass fuel sources using this technology is in its infancy. The plasma sources considered are typically directed plasma torches, with associated thermal losses from water cooling of the electrodes.[54] However, it is also possible to construct a gasifier using a plasma arc source maintained between uncooled graphite electrodes.[57] In contrast to most approaches to waste processing, which have used inert gases (usually argon or nitrogen) in the plasma torch and have not been unduly concerned with the gas flow rate from the torch, the production of high-quality syngas from plasma processing requires a minimal use of plasma gas (if a traditional inert gas is used) or else operation of the plasma torch on steam[58] or on a steam-syngas mixture.[59] Many designs, including the well-known Westinghouse plasma gasifier[60] and the Solena gasifier[61] use oxygen to assist in char conversion, but this entails a potentially significant cost, as discussed in the last section. When using an inert plasma gas, the flow of plasma gas is minimized by generating a very high temperature plasma and then mixing steam into the plasma column.[31,48,62,63]

The means by which the biomass feedstocks are coupled to the plasma source(s) vary significantly for different process configurations and have important implications for both the requisite preparation of the biomass feed and the quality of the syngas produced. Some approaches are based on direct-feed of the solid feedstock into the plasma stream, necessitating fairly small particle sizes to attain the requisite conversion in a limited plasma exposure time. This type of process requires fuel preparation that is similar to that required for entrained flow gasification. On the other hand, in both the Westinghouse plasma gasifier and the Solena gasifier the biomass is fed on top of a moving bed that has impinging plasma torches.[60,61] In this case, no special feed preparation is necessary, but the hot syngas flows upwards through the raw biomass feed, generating significant quantities of hydrocarbons and tar, similar to the indirect gasification process. Consequently, secondary tar cracking and/or reforming or tar capture and reinjection is necessary in this approach.

The thermal plasma gasification technology has only been developed for pressures near 1 atm., so significant compression of the syngas is necessary before liquid fuel synthesis can occur. On the other hand, the usable syngas yield (i.e. $CO + H_2$) for plasma gasification far exceeds that from the other gasification technologies, because in the other approaches some of the biomass feedstock is consumed (through combustion reactions) to provide the heat necessary to sustain the gasification process. Whether this primary advantage of plasma gasification can truly offset the cost of electricity consumed in generating the plasma has yet to be practically determined.

7.4   Gas Cleanup

Synthesis catalysts are sensitive to various impurities that may exist in the gas stream exiting the gasification process. These catalysts can be poisoned by ammonia (NH_3), hydrogen cyanide (HCN), hydrogen sulfide (H_2S), and carbonyl sulfide (COS). Hydrochloric acid (HCl) and hydrofluoric acid (HF) can corrode the catalysts, while CO_2 can inhibit the synthesis process. Hydrocarbons and tars result in catalyst coking, reducing catalyst activity over time. Particles, such as dust, soot, or alkali fume, will deposit on catalysts and reduce their effectiveness. Consequently, all of these contaminants must be reduced to acceptably low levels (typically from 10 ppb to 1 ppm, depending on the contaminant) before passing the syngas to the synthesis unit.[30,64–66]

Significant effort has been devoted to the development of hot-gas cleanup approaches that can remove all of the contaminants from gasifier syngas.[64,67–70] Removal of contaminants at elevated temperatures reduces heat transfer losses associated with syngas cooling and also eliminates the production of waste-water from the gas cleaning process. However, only some of the cleanup steps have been successfully applied at high temperatures, so current practice is to cool the syngas and conduct most of the gas cleaning at low temperatures.

The concentration of nitrogen- and sulfur-containing contaminants is primarily a function of the biomass feedstock and is fairly independent of the type of gasification technology that is employed. Chlorine species are also carried by the biomass source, and can vary widely, depending on the proximity of the biomass source to the ocean, for example, but the concentration of HCl in the syngas is also somewhat dependent on the gasification tech-nology. Moderate temperature bed-based processes tend to have lower HCl content in the syngas because a larger fraction of the chlorine binds to the ash.

The primary differences in syngas contaminants as a function of gasification technology lie in the areas of hydrocarbons and tar, dust loading, and alkali fume. The lower temperature processes produce much higher levels of hydro-carbons and tar, as has been previously discussed. Bed-based processes also tend to produce some fine dust in the product gas, particularly for circulating fluidized beds. High temperature gasification processes, such as entrained flow and thermal plasma gasifiers, produce alkali fume.

Figure 7.3 shows a schematic of a typical, multi-step biomass syngas cleanup process. The first step in gas cleanup usually addresses the syngas tar, for those gasification processes that produce a significant tar loading. This is because cooling the raw syngas below 400 °C will begin to condense out the tars, leading to wall deposits and downstream deposition issues. In some designs, the tars are removed by condensation (also removing the dust) and then reinjected into the gasifier.[29,61] In other designs, the hydrocarbons and tar are steam reformed in a catalyst bed. A large amount of research has been conducted on catalysts for tar reforming.[71] Most of this work has focused on nickel catalysts on an alumina substrate, often with a metallic promoter (such as Mg). These catalysts tend to be much more effective at reforming tars than at reforming hydrocarbons, particularly methane. As the sulfur has not been removed at this point in the clean-up process, sulfur poisoning of the tar reforming catalyst is a concern, in addition to catalyst coking. Research is ongoing to discover effective sulfur-tolerant, coke-inhibiting catalysts for tar reforming.

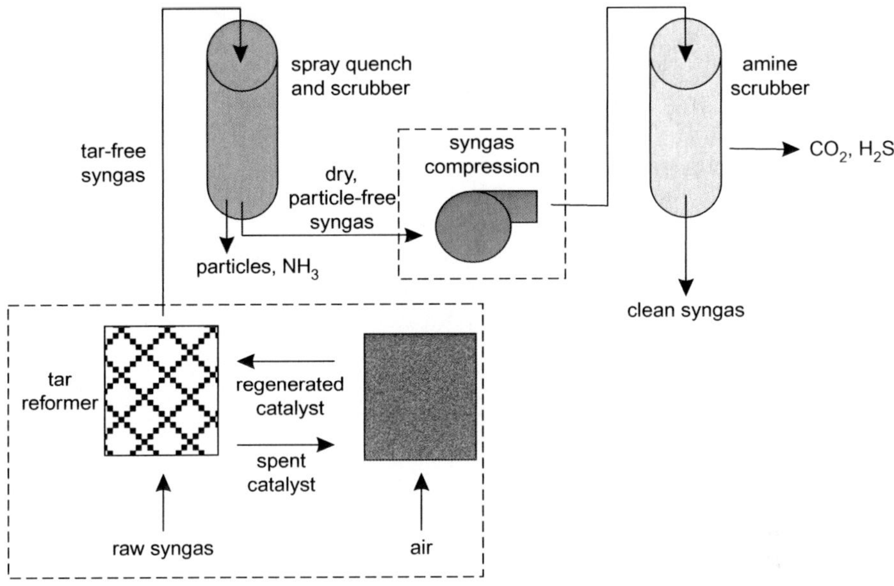

Figure 7.3 Schematic of a typical gas cleanup configuration for treating gasifier syngas before passing the syngas to a fuels synthesis unit. The process steps surrounded by dashed boxes are only required when treating the syngas from certain types of gasifiers, as described in the text.

After the tars have been removed, the syngas is dried and the entrained particles are removed through a syngas quench and wet scrubbing system. In most cases the alkali fume particles are also effectively captured in wet scrubbing systems. This process also removes any water-soluble gas species such as HCN, NH$_3$, HCl, and HF. The quench and wet scrubbing system necessarily entails the loss of much of the sensible heat of the hot syngas that enters the process. An alternative approach to capturing particles while maintaining elevated temperatures is to use hot barrier filters, such as ceramic candle filters,[64] but then other approaches are needed to remove the other species usually captured in wet scrubbing, such as the nitrogen-containing species, HCl, and HF.

Acid gas removal is strongly facilitated by operation at pressure. Therefore, for those gasifier designs operating near 1 atmosphere, it is useful to compress the syngas that has otherwise been cleaned and is suitable for use within a compressor. Gasifier designs that can operate at pressure avoid this costly step. Once compressed, the acid gases can be removed, either together, as suggested in Figure 7.3, or in series operations. One of the most common means of acid gas removal for biomass gasification systems is to employ an amine scrubber, using a solvent such as monoethanolamine (MEA). This solvent effectively adsorbs both H$_2$S and CO$_2$ at a temperature of approximately 40 °C and then releases these gases when heated in an adjacent regenerator vessel to 120 °C. For hot-gas cleanup, metal oxide systems, particularly involving zinc oxide (ZnO), have been shown to be effective for capture of sulfur-containing species.[29,70]

7.5 Conclusions

Gasification is a very flexible process for utilizing a wide variety of biomass feedstocks and generating any number of gaseous or liquid fuels, including synthetic natural gas, hydrogen, methanol, ethanol, gasoline, diesel, or DME. Historically, most operating experience for biomass gasification has either involved air-blown gasification, producing a highly diluted product gas that is unsuitable for fuel synthesis, or steam-blown indirect (allothermal) gasification.

The allothermal gasification process has been actively investigated by a number of researchers and technology developers over the years and has been scaled up to a size that can process 200 tpd. Unfortunately, this process produces large quantities of tars and hydrocarbons, requiring secondary reforming of the produced syngas, and thus far has only been operated at low pressures. Recent advances in this technology have centered on improved thermal integration of the steam gasification/char combustion chambers, to reduce thermal losses in the system. There is also a significant amount of research currently focused on the development of effective tar reforming catalysts. If cost-effective catalysts can be developed that reform small hydrocarbon species, in addition to tars, and which are sulfur-tolerant, the allothermal gasification approach will be much more promising for syngas applications. However, the low-pressure nature of this process, as currently implemented, still creates significant limitations to its use for liquid fuel synthesis.

Direct (autothermal) gasification of biomass has only been commercially employed to-date as a co-feed with coal or petroleum coke into large-scale entrained flow coal gasifiers. The capital and operating costs of oxygen production, especially at the moderate scales required for dedicated biomass gasifiers, have historically restricted the introduction of this technology, though it has the distinct advantage, for purposes of liquid fuel synthesis, of easily being employed at elevated pressure. Also, high temperature entrained flow gasifiers produce negligible char. When operating at pressure, the main difficulty stems from processing the biomass source to a size and consistency to allow feeding against the pressure gradient. Small fuel particle sizes are also required to assure sufficient conversion in entrained flow gasifiers. Recent improvements in production of oxygen via pressure swing adsorption (PSA), especially the implementation of the vacuum PSA technique, offer the promise of lower oxygen production costs at the appropriate scale. Such reductions in oxygen production costs may make autothermal gasification, specifically in the form of entrained flow gasifiers, the preferred gasification route for biomass for liquid fuel synthesis (as it already is for coal gasification).

Thermal plasma gasification was originally developed for application to waste streams, with the high temperature plasma source guaranteeing complete destruction of the waste. As applied to biomass gasification for liquid fuel synthesis, plasma gasification offers the advantage of higher yields of fuels from the biomass source than any other type of gasification approach. However, significant constraints to the practical implementation of plasma gasification lie in the high cost of electricity use (in the plasma source), oxygen use (if needed), electrode replacement, and subsequent syngas compression (because of low-pressure operation of the plasma sources). Furthermore, some designs create significant hydrocarbon and tar loading, necessitating subsequent syngas reforming.

References

1. K. H. Casleton, R. W. Breault and G. A. Richards, *Combust. Sci. Technol.*, 2008, **180**, 1013.
2. M. Molieri, *Int. J. Thermal Sci.*, 2000, **39**, 141.
3. A. Campbell, J. Goldmeer, T. Healy, R. Washam, M. Molieri and J. Citeno, *Proc. ASME Turbo Expo 2008*, 2008, **3**, 1077.
4. T. Seitarides, C. Athanasiou and A. Zabanlotou, *Renew. Sustain. Energy Rev.*, 2008, **12**, 1251.
5. S. Turn, C. Kinoshita, Z. Zhang, D. Ishimura and J. Zhou, *Int. J. Hydrogen Energy*, 1998, **23**, 641.
6. R. M. Navarro, M. A. Peña and J. L. G. Fierro, *Chem. Rev.*, 2007, **107**, 3952.
7. C. M. White, R. R. Steeper and A. E. Lutz, *Int. J. Hydrogen Energy*, 2006, **31**, 1292.
8. C. C. Cormos, *Int. J. Hydrogen Energy*, 2009, **34**, 6065.
9. S. Verhelst and T. Wallner, *Prog. Energy Combust. Sci.*, 2009, **35**, 490.
10. G. Frenette and D. Fornhoffer, *Int. J. Hydrogen Energy*, 2009, **34**, 3578.
11. R. Toonssen, N. Woudstra and A. H. M. Verkooijen, *J. Power Sources*, 2009, **194**, 456.
12. P. Corbo, F. Migliardini and O. Veneri, *Energy Conv. Mgmt.*, 2007, **48**, 2365.
13. P. Chiesa, G. Lozza and L. Mazzocchi, *J. Eng. Gas Turbines Power*, 2005, **127**, 73.
14. C. M. van der Meijden, H. J. Veringa, B. J. Vreugdenhil and B. van der Drift, *Int. J. Chem. Reactor Eng.*, 2009, **7**, A53.
15. P. J. A. Tijm, F. J. Waller and D. M. Brown, *Appl. Cat. A – General*, 2001, **221**, 275.
16. F. Gallucci and A. Basile, *Int. J. Hydrogen Energy*, 2007, **32**, 5050.
17. F. J. Keil, *Microporous Mesoporous Mater.*, 1999, **29**, 49.
18. T. Mokrani and M. Scurrell, *Catal. Rev.*, 2009, **51**, 1.
19. F. Raloof, M. Taghizadeh, A. Eliassi and F. Yaripour, *Fuel*, 2008, **87**, 2967.
20. X. D. Peng, A. W. Wang, B. A. Toseland and P. J. A. Tijm, *Ind. Eng. Chem. Res.*, 1999, **38**, 4381.
21. Y. Adachi, M. Komoto, I. Watanabe, Y. Ohno and K. Fujimoto, *Fuel*, 2000, **79**, 229.
22. Y. P. Li, T. J. Wang, X. L. Yin, C. Z. Wu, L. L. Ma, H. B. Li, Y. X. Lu and L. Sun, *Renewable Energy*, 2010, **35**, 583.
23. S. C. Sorensen, *J. Eng. Gas Turb. Power*, 2001, **123**, 652.

24. T. A. Semelsberger, R. L. Borup and H. L. Green, *J. Power Sources*, 2006, **156**, 497.
25. M. C. Lee, S. B. Seo, J. H. Chung, Y. J. Joo and D. H. Ahn, *Fuel*, 2009, **88**, 657.
26. A. Bharadwaj, L. L. Baxter and A. L. Robinson, *Energy & Fuels*, 2004, **18**, 1021.
27. J. Corella, J. M. Toledo and G. Molina, *Ind. Eng. Chem. Res.*, 2007, **46**, 6831.
28. J. J. Hernandez, G. Aranda-Almansa and A. Bula, *Fuel Proc. Technol.*, 2010, **91**, 681.
29. C. M. van der Meijden, H. J. Veringa and L. P. L. M. Rabou, *Biomass Bioenergy*, 2010, **34**, 302.
30. W. Zhang, *Fuel Proc. Technol.*, 2010, **91**, 866.
31. M. Hrabovsky, M. Hlina, M. Konrad, V. Kopecky, T. Kavka, O. Chumak and A. Maslani, *High Temp. Mater. Proc.*, 2009, **12**, 299.
32. D. C. Elliott, *Biofuels, Bioprod. Bioref.*, 2008, **2**, 254.
33. Y. Wu, S. Wu, Y. Li and J. Gao, *Energy Fuels*, 2009, **23**, 5144.
34. S. Xiong, M. Öhman, Y. Zhang and T. Lestander, *Energy Fuels*, 2010, **24**, 4866.
35. Y. Niu, H. Tan, X. Wang, Z. Liu, H. Liu, Y. Liu and T. Xu, *Biores. Technol.*, 2010, **101**, 9373.
36. C.-L. Hsi, T.-Y. Wang, C.-H. Tsai, C.-Y. Chang, C.-H. Liu, Y.-C. Chang and J.-T. Kuo, *Energy & Fuels*, 2008, **22**, 4196.
37. Catalytic Cracking, in *The Chemistry and Technology of Petroleum*, 4th edn, ed. J. G. Speight, CRC Press, Boca Raton, 2007.
38. M. A. Paisley, M. C. Farris, J. W. Black, J. M. Irving and R. P. Overend, 'Preliminary operating results from the Battelle/FERCO gasification demonstration plant in Burlington, Vermont, USA,' in *1st World Conference on Biomass for Energy and Industry*, ed. S. Kyritsis and A. A. C. M. Beenackers, James & James (Science Publishers), Ltd., London, 2001, Vol. 2, pp. 1494–1497.
39. T. Murakami, G. W. Xu, T. Suda, Y. Matsuzawa, H. Tani and T. Fujimori, *Fuel*, 2007, **86**, 244.
40. K. Matsuoka, K. Kuramoto, T. Murakami and Y. Suzuki, *Energy & Fuels*, 2008, **22**, 1980.
41. C. Pfeifer, R. Rauch and H. Hofbauer, *Ind. Eng. Chem. Res.*, 2004, **43**, 1634.
42. G. W. Xu, T. Murakami, T. Suda, Y. Matsuzawa and H. Tani, *Ind. Eng. Chem. Res.*, 2006, **45**, 2281.
43. G. W. Xu, T. Murakami, T. Suda, Y. Matsuzawa and H. Tani, *Fuel Proc. Technol.*, 2009, **90**, 137.
44. R. L. Bain, 'Material and energy balances for methanol from biomass using biomass gasifiers,' NREL Technical Report, NREL/TP-510-17098, January 1992.
45. P. Spath, A. Aden, T. Eggeman, M. Ringer, B. Wallace and J. Jechura, 'Biomass to hydrogen production detailed design and economics utilizing the

Battelle Columbus Laboratory indirectly-heated gasifier,' NREL Technical Report, NREL/TP-510-37408, May 2005.

46. X. B. Xiao, D. D. Le, K. Morishita, S. Zhang, L. Li and T. Takarada, *Fuel Proc. Technol.*, 2010, **91**, 895.
47. I. Iliuta, A. Leclerc and F. Larachi, *Biores. Technol.*, 2010, **101**, 3194.
48. A. J. Minchener, *Fuel*, 2005, **84**, 2222.
49. W. He, C. S. Park and J. M. Norbeck, *Energy Fuels*, 2009, **23**, 4763.
50. M. J. Prins, K. J. Ptasinski and F. J. J. G. Janssen, *Energy*, 2006, **31**, 3458.
51. J. C. Santos, P. Cruz, T. Regala, F. D. Magalhães and A. Mendes, *Ind. Eng. Chem. Res.*, 2007, **46**, 591.
52. H. Nishikawa, M. Ibe, M. Tanaka, M. Ushio, T. Takemoto, K. Tanaka, N. Tanahashi and T. Ito, *Vacuum*, 2004, **73**, 589.
53. H. Huang and L. Tang, *Energy Conversion Mgmt.*, 2007, **48**, 1331.
54. E. Gomez, D. Amutha Rani, C. R. Cheeseman, D. Deegan, M. Wise and A. R. Boccaccini, *J. Haz. Mater.*, 2009, **161**, 614.
55. L. Tang and H. Huang, *Fuel*, 2005, **84**, 2055.
56. W.-K. Tu, J.-L. Shie, C.-Y. Chang, C.-F. Chang, C.-F. Lin, S.-Y. Yang, J.-T. Kuo, D.-G. Shaw and D.-J. Lee, *Energy & Fuels*, 2008, **22**, 24.
57. K. Moustakas, D. Fatta, S. Malamis, K. Haralambous and M. Loizidou, *J. Haz. Mater.*, 2005, **B123**, 120.
58. H. S. Park, C. G. Kim and S. J. Kim, *J. Ind. Eng. Chem.*, 2006, **12**, 216.
59. M. Brothier, P. Gramondi, C. Poletiko, U. Michon, M. Labrot and A. Hacala, *High Temp. Mat. Proc.*, 2007, **11**, 231.
60. http://www.westinghouse-plasma.com/technology/what-is-plasma-gasification, accessed Nov. 29, 2010.
61. http://www.solenagroup.com/techfiles/pgv-2010.html, accessed Nov. 29, 2010.
62. L. Tang, H. Huang, A. Ahao, C. Z. Wu and Y. Chen, *Ind. Eng. Chem. Res.*, 2003, **42**, 1145.
63. H. Nishikawa, M. Ibe, M. Tanaka, T. Takemoto and M. Ushio, *Vacuum*, 2006, **80**, 1311.
64. D. J. Stevens, 'Hot gas conditioning: recent progress with larger-scale biomass gasification systems,' NREL Summary Report, NREL/SR-510-29952, August 2001.
65. S. D. Phillips, *Ind. Eng. Chem. Res.*, 2007, **46**, 8887.
66. A. Kumar, D. D. Jones and M. A. Hanna, *Energies*, 2009, **2**, 556.
67. W. Torres, S. S. Pansare and J. G. Goodwin, Jr., *Catal. Rev.*, 2007, **49**, 407.
68. H. Leibold, A. Hornung and H. Seifert, *Powder Technol.*, 2008, **180**, 265.
69. C. Xu, J. Donald, E. Byambajav and Y. Ohtsuka, *Fuel*, 2010, **89**, 1784.
70. X. M. Meng, W. de Jong, R. Pal and A. H. M. Verkooijen, *Fuel Proc. Technol.*, 2010, **91**, 964.
71. M. W. Yung, W. S. Jablonski and K. A. Magrini-Bair, *Energy & Fuels*, 2009, **23**, 1874.

Bioinspired Catalysts for Biofuels: Challenges and Future Directions

TED J. AMUNDSEN AND ALEXANDER KATZ

Department of Chemical and Biomolecular Engineering MC 1462, 110A
Gilman Hall, University of California at Berkeley, Berkeley, California

8.1 Introduction

Enzymes implement multifunctional combinations of binding events and
catalytically active sites to orchestrate reaction pathways, which often have no
counterpart in synthetic catalyst systems lacking the exquisite microenviron-
ment found in the enzyme.[1] Cellulose hydrolysis represents a prime example of
a transformation that is at the heart of biofuels processing and that is also one
that enzymes catalyze with higher selectivity relative to what is achievable with
synthetic catalysts. Glycosidases are considered a prime example of a proficient
of enzyme[2] and in particular engender glycosidic bond hydrolysis rate
enhancements of up to 10^{17}-fold over background in solution.[3] Glycosidases are
thought to employ acid–base bifunctional catalysis in the form of a carboxylic
acid–carboxylate pair at the enzymatic active site. Early research identified this
functional group pairing at the active site of hydrolytic enzymes.[4,5] The cellulose
hydrolysis enzyme lysozyme is one of the first to be characterized with these
paired functional groups in its catalytically active site, consisting of Glu and Asp
residues as acid and conjugate base, respectively.[6] The cooperative action of these

RSC Energy and Environment Series No. 4
Chemical and Biochemical Catalysis for Next Generation Biofuels
Edited by Blake Simmons
© Royal Society of Chemistry 2011
Published by the Royal Society of Chemistry, www.rsc.org

two residues results in either retention or inversion of anomeric configuration depending on the relative distances between the reactive functionalities.[2]

A tantalizing and open-ended question is: what is the minimum amount of complexity that is necessary to incorporate into a synthetic catalyst in order to mimic some of the most quintessential features of enzymes? Answers to this question can lead to bioinspired catalysts that shed further light on the mechanism of enzymes, by providing insight into the relative contributions of various factors suspected to influence catalytic activity, and aid in elucidating the requirements for each of these contributors to be successful. Even a sophisticated bioinspired catalyst is relatively simple compared to the nonlinear and dynamic structural complexity found in an enzyme. It is therefore much easier to manipulate different components of the bioinspired catalyst than it is to selectively impart rational changes to enzymes. In addition, and perhaps more usefully, bioinspired catalysts provide a basis for the development of synthetic catalysts, and typically have a combination of favorable character-istics of synthetic catalysts such as stability, cost, recyclability, tunability, etc., when compared with enzymes. Such numerous advantages of synthetic cata-lysts, either homogeneous or heterogeneous, make their implementation fea-sible even if the targeted enzymatic action is only crudely reproduced. See for example the acid–base bifunctional catalysts consisting of a tethered primary amine and acid on silica as synthesized by Davis *et al.*,[7] which have been nicely reviewed alongside other heterogeneous bifunctional catalysts.[8] These catalysts are discussed below as highly active bifunctional catalysts for aldol con-densation reactions. Therefore, catalyst design guided by invoked mechanisms of biological systems represents fertile ground for the development of selective catalysts capable of performing many highly desired transformations, such as those ubiquitous in biofuels production. Below, some of the essential features of enzymes are reviewed and applied to chemical transformations of relevance to biofuels processing: acyl transfer, ester hydrolysis, glycosidic bond hydrolysis, aldol condensation, ketonization, dehydration, and lignin depolymerization.

8.2 Substrate Binding

Substrate binding is suspected to account for a significant portion of enzymatic catalytic activity, by effectively creating a large local concentration of the catalytically active moieties that are positioned in proximity to the binding site.[8] The loss of translational and rotational entropy upon binding has been previously hypothesized to account for rate accelerations of up to 10^8 by comparing rates of intramolecular and intermolecular reactions.[9] Binding can also direct the substrate into a particular orientation within the active site, which allows the enzyme to both achieve a high degree of selectivity in its catalytic transformations and, when necessary, distort the conformation of substrates into more reactive geometries. The latter is apparent in the reactions of chymotrypsin where a nitrogenous oxyanion hole is thought to facilitate the formation of a kinetically relevant tetrahedral intermediate,[10,11] and in

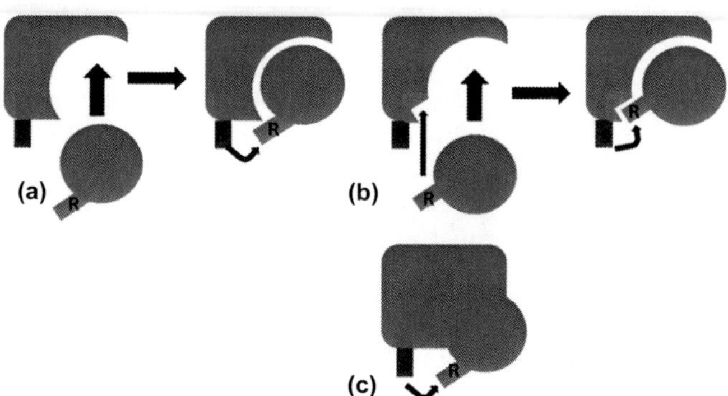

Figure 8.1 Schematic representing a catalyst active site (S) interacting with a substrate reactive center (R) during (a) unactivated binding, (b) activated binding, and (c) intramolecularly.

glycosidases, where distortion of hexose substrates into the more reactive, and carbenium ion intermediate-like, half-chair conformation is invoked.[12]

Binding has been incorporated into several bioinspired synthetic systems.[13] Cyclodextrins,[14] for instance, have been functionalized with catalytically active groups in order to exploit their hydrophobic interior as a reaction medium, which provides a binding pocket for non-polar substrates and can more effectively solvate these molecules than water or aqueous solution. This type of system is shown in Figure 8.1a. It is bifunctional in the sense that there is a binding event taking place along with catalysis. In addition to binding–catalysis bifunctionality, which will be referred to as unactivated binding, there can be a situation where the binding interaction between the catalyst and the substrate actually creates a new reaction pathway not possible with only one of the functionalities alone. This system is bifunctional in more than one sense because in addition to binding–catalysis bifunctionality, there is a cooperative bifunctionality between catalytic groups which function in concert to facilitate catalytic transformations. This type of binding is referred to as activated binding, and is schematically illustrated in Figure 8.1b. These basic constructs complete the framework for describing general modes of action of bioinspired catalysts.

8.2.1 Competitive Aqueous Solvation

A bifunctional bioinspired catalyst should ideally be able to successfully compete against monofunctional catalysts operating in water.[15] This is especially important in biomass to biofuels conversions where the cost of completely dehydrating biomass is prohibitive. However, in practice, this is difficult to accomplish. Take, for instance, the performance of one of the earliest reported cooperative catalyst systems, Swain and Brown's 2-hydroxypyridine for the mutarotation of tetramethylglucose.[16] The reaction is second order in the presence of the bifunctional 2-hydroxypyridine catalyst while it is third order when a combination of monofunctional acids and bases are used. Furthermore, the bifunctionally catalyzed reaction is 7000 times faster than that catalyzed by a mixture of monofunctional acid and base of similar strength under identically dilute conditions. These results pertain to the reaction conducted in benzene where there are no competitive hydrogen bond donors. When conducted in water, however, 2-hydroxypryidine is if at all, only marginally more effective than monofunctional catalysts. Water is a good solvent for polar molecules because it readily donates or accepts protons as solvent, and the large concentration of water in aqueous solution, 55 M, makes it difficult for synthetic molecular receptors present at much lower concentrations to compete. This can be illustrated by the extreme example of water binding in the (very) hydrophobic region consisting of the shallow interior of a calixarene cavity. Water effectively solvates this cavity with a free energy of adsorption that is nearly half of the free energy of vaporization. This has the effect of decreasing binding constants of organic molecules from aqueous solution relative to vapor phase (where competitive adsorption of water is not an issue) by an order of magnitude.[17] This screening effect of water is anticipated to be even more apparent in systems implementing polar binding sites, such as hydrogen bonding molecular receptors. The comparison between solvation by either a receptor molecule for activated binding of substrate or water in aqueous solution is akin to that of inter- and intra-molecular catalysis shown in Figures 8.1b and 8.1c. The latter has a distinct advantage over the former because an intramolecular system forces the substrate into what can essentially be viewed as a 100% bound state.

8.2.2 Positional Requirements

Bioinspired catalysts often require precise positional requirements for achieving bifunctional catalysis. This is perhaps most clearly evident in examples of synthetic bioinspired catalysts consisting of cyclodextrins. Substituted cyclodextrins in particular have been implemented as catalysts for different transformations,[14] including glycosidic bond hydrolysis, discussed in more detail below. While the interior of the cyclodextrin effectively binds non-polar substrates, catalytic groups can be appended to either the primary or secondary face or both typically via ester, amide, ether, or disulfide bonds. Adjustments on the order of a few Ångstroms are enough to often make the difference between activity and lack of activity for a cyclodextrin-based bioinspired catalyst. This is exemplified by the dramatic catalytic consequences of the positioning of an imidazole moiety relative to the hydrophobic binding pocket of a functionalized cyclodextrin catalyst. Cramer *et al.* first synthesized an artificial chymotrypsin mimic consisting of an imidazole functionalized β-cyclodextrin which serves as an unactivated binding agent.[18] This molecule has the imidazole substituted onto the primary face of the cyclodextrin, facing away from the hydrophobic binding pocket. The largest rate enhancement attained with these functionalized cyclodextrins over similarly substituted imidazoles lacking any cyclodextrin functionality is a factor of 7. Several years later, as the result of a novel synthetic technique, Rao *et al.* synthesized a β-cyclodextrin

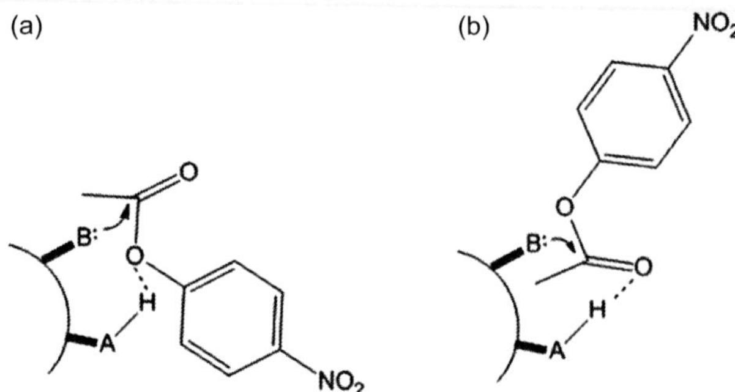

Figure 8.2 Productive (a) and non-productive (b) proton donation during bifunctional catalysis of *p*-nitrophenyl acetate hydrolysis.

functionalized with imidazole at the secondary face, pointing towards the interior of the binding pocket.[19] As expected, the hydrolytic performance of this enzyme mimic was greatly enhanced over its predecessor, exhibiting a 70-fold improvement in rate of *p*-nitrophenyl acetate (*p*NPA) hydrolysis. Some of this improvement may also be due to the shorter, more rigid tether connecting the imidazole to the cyclodextrin in the case of the Rao *et al.* system *versus* the Cramer *et al.* system. These examples highlight the importance of organizational effects on bioinspired catalyst function, for a bifunctional bioinspired catalyst consisting of binding and tethered catalytic elements. Furthermore, these synthetic challenges can persist in even more subtle ways, such as in the activated binding of the bifunctional ester hydrolysis catalysts described below. While an enzyme can ensure that a catalytically relevant acid will be positioned optimally for catalysis, more simplistic bioinspired catalysts have difficulty directing the acid to interact in the most effective fashion, given multiple possible configurations (Figure 8.2).

8.3 Acyl Transfer

Acyl transfer and ester hydrolysis are relevant to transformations involving biodiesel production. The mechanistically related transesterification is the most widely used reaction to synthesize biodiesel,[20] and can occur via similar reaction pathways to those described in the following sections.[21] Acyl transfer and ester hydrolysis can be implemented for hydrolysis of ester feedstocks where products formed are subsequently esterified.[22]

8.3.1 Hydrogen Bonding Molecular Receptors

Tecilla and Hamilton elegantly synthesized an acyl transfer catalyst consisting of groups capable of either donation or acceptance of hydrogen bonds with a substrate molecule.[23,24] Thiolated molecular receptor catalysts form various numbers of hydrogen bonds with barbiturate derivative substrates during acyl

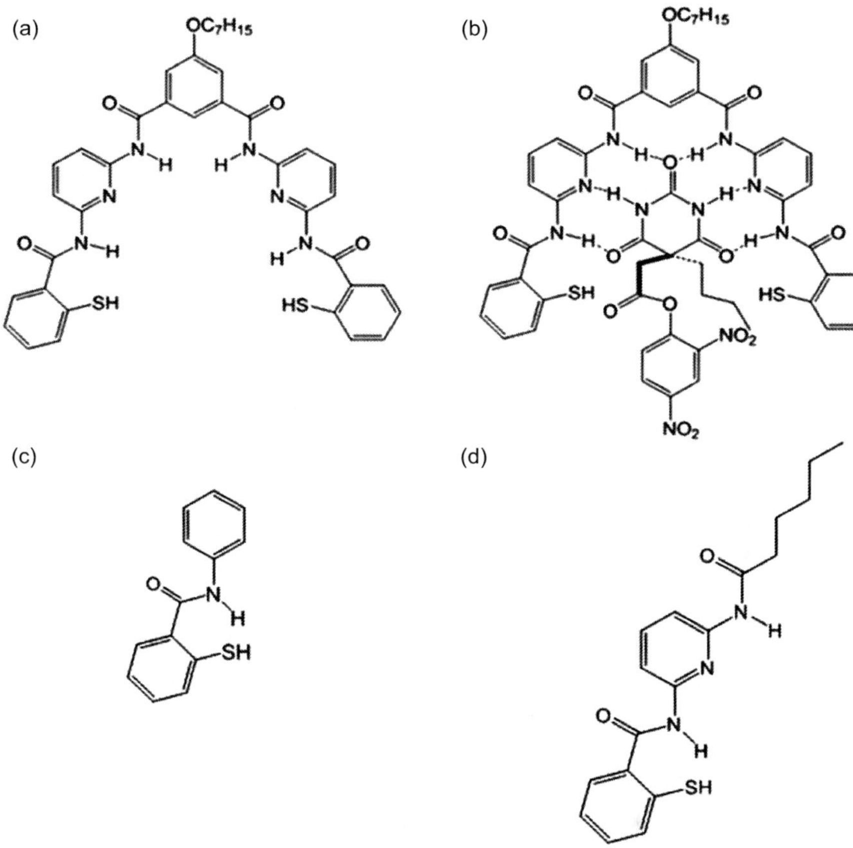

Figure 8.3 Molecular receptors (a), (c), and (d) capable of participating in unactivated binding (b) with a barbiturate ester during acyl transfer to the receptor thiol group (adapted from ref. 25).

transfer, as shown in Figure 8.3b.[25] The receptor with the most points of interaction with the substrate in their system, corresponding to Figure 8.3a, produces up to 10^4 fold rate enhancements over less complex analogs, which bound the substrates with fewer hydrogen bonds, as shown in Figures 8.3c and 8.3d. The catalytic mode of action of these intricate receptors appears to be via unactivated binding where the ester carbonyl is brought into proximity of the thiol where acyl transfer can readily occur. These reactions were performed in dichloromethane, where water is not present to compete with the receptor portion of the catalyst, and would likely perform more poorly in water due to competitive binding.

In order to draw larger distinctions between bifunctionally catalyzed reactions and monofunctional analogs in competitive polar solvents such as water, bioinspired catalysts must rely on activated binding. Like the example of Swain and Brown,[16] activated binding involves a favorable interaction between catalyst and substrate at the actual active site. Hamilton's group later synthesized another hydrogen bonding molecular receptor that engaged in activated binding with the substrate. It is able to function in acetonitrile (a much more polar solvent), in which the catalysts based on unactivated binding are unable to demonstrate enhanced rates of reaction. In this case the receptor donated protons to the phosphate diester substrate in such as way as to increase the electrophilicity of the phosphorous, and in doing so increased the trans-esterification rate to 290 times that observed over background reaction.[26]

For instance, in a retaining glycosidase, acid–base bifunctional catalysis is suspected to occur through the simultaneous protonation of the leaving alcohol along with the nucleophilic addition of the carboxylate base to form a covalent intermediate.[2] Activated binding is able to influence the reaction rate directly, not only by a local concentration effect as in the case of unactivated binding, thus drawing a greater distinction from analogous monofunctional counterparts even in aqueous environments. Take for example the aldol reactions discussed below. It is generally accepted that these reactions can be catalyzed bifunctionally, where proton donation from a catalytic acid to the basic carbonyl oxygen of the substrate activates the carbonyl towards nucleophilic attack. In this case, the strength of the acid will determine the extent of carbonyl activation. Therefore, if an acid stronger than water is positioned in such as way as to act in concert with a catalytic base, then the reaction should be able to proceed via a pathway that would not exist in water alone.

Finally, another effective method to produce bifunctionally catalyzed rates that are much larger than similar monofunctionally catalyzed reactions in water is to simply use a higher concentration of catalyst. This shifts the binding equilibrium towards the bound substrate and subsequently results in a proportionally larger rate as the fraction of substrate bound increases. Increasing the amount bound aids in offsetting the large advantage water has from its high 55 M concentration, and facilitates observing reaction rates above and beyond any background reaction in water.

8.3.2   Aqueous Hydrogen Bonding Molecular Receptors

Motomura *et al.* implemented the above-mentioned techniques and found that the aqueous deacylation of *p*NPA was catalyzed at a much faster rate by a molecule containing both imidazole and bisresorcinol moieties (Figure 8.4) than by either of these functionalities alone, or by a mixture of these molecules.[27] Specifically, acyl transfer from *p*NPA to the imidazole moiety of the molecule containing resorcinol is 20-fold faster than the same reaction with an analogous imidazole derivative lacking any acidic functionality, but maintaining the same imidazole pK_a. This is all the more impressive given the fact that (i) the less reactive monofunctional catalyst is clearly less sterically crowded than the bifunctional counterpart and (ii) steric congestion at the active site is expected to have a negative influence on reaction rates.[28] Bifunctional catalysis in this system is expected to result from activation of the carbonyl substrate which facilitates rate-limiting addition of imidazole.[29]

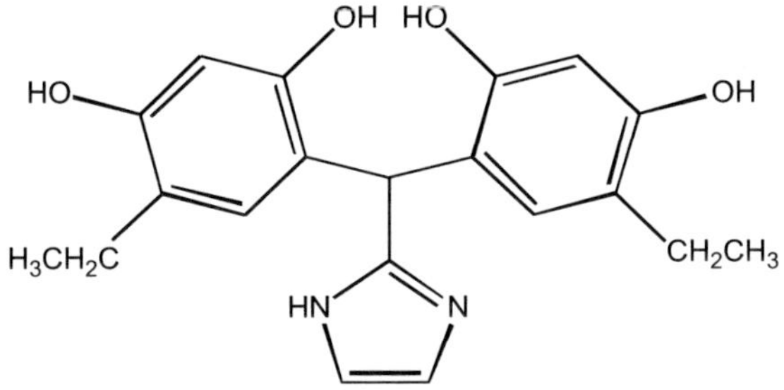

Figure 8.4 Imidazole–bisresorcinol derivative used as bifunctional acyl transfer catalyst (adapted from ref. 27).

Molecular modeling confirms[27] that the geometry of the tetrahedral intermediate is such that there can be proton donation from the acidic phenols of the bisresorcinol derivative to the carbonyl oxygen of the *p*NPA substrate.

These results are interesting for several reasons. First, the bifunctional effects are present in water. Second, this system appears capable of implementing activated binding to bring about some of the rate enhancements. The binding could potentially create a new reaction pathway by activating the substrate to nucleophilic attack. Lastly, this reaction was conducted in a regime where there is a 50-fold excess of catalyst relative to substrate, which serves to shift the equilibrium towards the bound substrate, and an exaggerated distinction can be drawn between the bifunctionally catalyzed reaction and any competitive non-specific contributions from water.

8.4 Ester Hydrolysis

Ester hydrolysis can be catalyzed by either acid or base, but the predominant pathway at mild pH is through base catalysis.[21] Within the realm of base catalyzed ester hydrolysis, there are two possible mechanisms: nucleophilic or general base catalysis. Nucleophilic catalysis, also known as covalent catalysis, refers to a reaction where the base catalyst forms a covalent bond with the substrate. In the case of ester hydrolysis, the bond is formed between the nucleophile and the carbonyl carbon of the substrate, creating a tetrahedral intermediate structure.[30] General base catalysis of ester hydrolysis refers to a mechanism where a base partially abstracts a proton from a water molecule, activating it as a nucleophile which then attacks the substrate carbonyl carbon.[31]

Butler and Gold performed a comprehensive study elucidating the relative contributions of both of these mechanisms in the hydrolysis of a series of activated aryl esters.[32] Their results demonstrate that the dominant mechanism is determined by the relative basicities of the leaving group alcohol and the attacking nucleophile. When the catalyst is less basic than the leaving group, the tetrahedral intermediate will partition primarily back to reactants, and general base catalysis is observed. This is because in such a system the tetrahedral intermediate will only partition to products when hydroxide ion is the nucleophile, a situation facilitated by the presence of a basic catalyst. Nucleophilic catalysis occurs when the catalyst is more basic than the leaving group alcohol, so that once an intermediate is formed, it is most likely to expel the leaving group and break down to products. Though often invoked to predict rate-limiting steps universally for ester hydrolysis reactions, below we demonstrate a caveat that the results of Butler and Gold do not apply so literally for bifunctional catalytic systems (see below).

In base catalyzed ester hydrolysis, a bifunctional mechanism consists of one where an acidic group on the catalyst either activates the substrate to nucleophilic attack by donating a proton to the carbonyl oxygen of the substrate, or facilitates the expulsion of the leaving group alcohol after the intermediate is formed by donating a proton to the leaving group oxygen. The mechanistic insight derived from Butler's work is relevant to bifunctional catalysis because it is far less likely for bifunctionality to be observed without the formation of a covalent intermediate between catalyst and substrate. If only serving to activate a water molecule, the base can act at a variable distance from the substrate due to very efficient aqueous proton conductivity (Figure 8.5b). With this type of mechanism, it is improbable that the painstakingly crafted catalyst will be in the proper orientation to induce a bifunctional effect because of the large number of possible relative orientations of catalyst to substrate. However, when there is a covalent bond formed between catalyst and substrate (Figure 8.5a), an acidic functionality would have a much higher probability of being in the proper orientation due to the requisite proximity of catalyst and substrate.

However, there is a caveat to the results of Butler and Gold when applied to bifunctional catalysis. If an acid is able to donate a proton to the leaving group oxygen during nucleophilic attack, it should be able to in effect lower leaving group basicity and therefore enhance leaving ability. This way, the contribution of nucleophilic catalysis to the overall reaction could in fact be much higher than predicted by the pK_a values of the basic catalyst and the leaving group alone (i.e. using data of Butler and Gold).

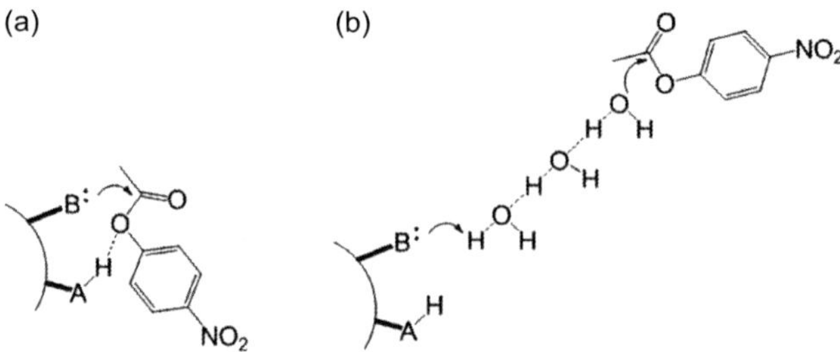

Figure 8.5   Acid–base bifunctional catalysis of nucelophilic *p*NPA hydrolysis (a) and same catalyst acting monofunctionally as a general base at variable distance (b).

8.4.1 Aqueous Hydrogen Bonding Molecular Receptors

The hypothesis above regarding bifunctional ester hydrolysis gains support through the results of Shaskus and Haake. They observed higher *p*NPA hydrolysis activity than predicted by its pK_a of 4.17 during aqueous catalysis by ascorbic acid monoanion.[33] Molecular modeling experiments show that the hydroxyl group in the 2 position, with a pK_a of 11.79, is able to donate its proton to the alcoholic leaving group of the tetrahedral intermediate formed by the addition of the ascorbate nucleophile to the carbonyl carbon of *p*NPA. As described above, this interaction is expected to accelerate the expulsion of the alcoholic leaving group from the tetrahedral intermediate, which is hypothesized to be the rate-limiting step when the leaving group is more basic than the attacking nucleophile. This is the case with ascorbate monoanion, p$K_a = 4.17$, and *p*NPA, p$K_a = 7.16$. As shown in Figure 8.6, the activity of ascorbate monoanion in *p*NPA hydrolysis is approximately 300 times faster than predicted by its pK_a based on the reactivity of other anionic nucleophiles in aqueous solution. In other words, ascorbate, with a pK_a of 4.17, is as reactive as a monofunctional oxyanionic nucleophile with a pK_a of approximately 6.7, even in the presence of competitive hydrogen bonding with water.

According to the results of Butler and Gold, *p*NPA hydrolysis by acetate anion, pK_a of 4.76, occurs only 46% through nucleophilic catalysis with the balance occurring through a general base mechanism.[32] If a similar figure is true of the ascorbate monoanion catalyzed reaction, then bifunctional catalysis would only be occurring in a fraction of the overall catalyzed rate, i.e. that which is going through a nucleophilic route. If this is correct, we would expect the hydrolysis of an ester with a less basic leaving group, such as

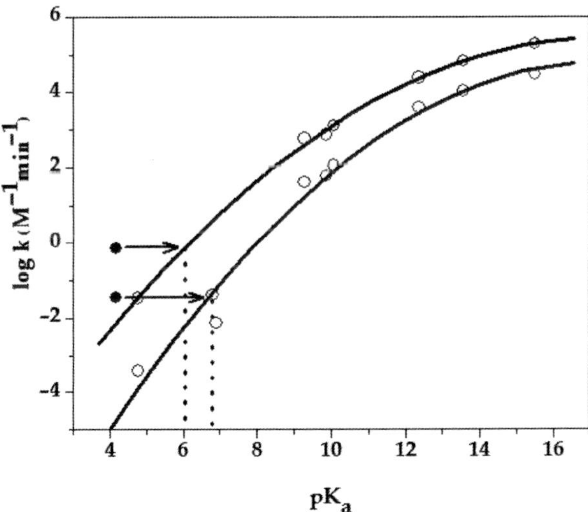

Figure 8.6 Hydrolysis of *p*NPA (bottom) and DNPA (top) by monofunctional anionic nucleophiles (open circles) of varying pK_a and ascorbic acid (filled circles). (Data from ref. 33.)

2,4-dinitrophenyl acetate (DNPA), which was found to be catalyzed by acetate via a nucleophilic pathway 82% of the time,[32] to exhibit an even larger rate enhancement than that of *p*NPA hydrolysis when subject to the bifunctional catalysis of ascorbate monoanion. However, the ascorbate catalyzed hydrolysis of DNPA, while still more efficient than predicted based on pK_a (Figure 8.6), exhibits less of an effect than in the hydrolysis of *p*NPA.[33] This observation can be attributed to several factors. As discussed above, it is quite possible that the relative contributions of nucleophilic and general base catalysis of aryl esters by anionic nucleophiles is different in the case of bifunctionally catalyzed systems, relative to what is predicted for monofunctional systems by results of Butler and Gold. The proton donation from the acid of the bifunctional catalyst to the oxygen of the alcoholic leaving group makes the leaving group less basic, and facilitates its expulsion. Therefore, the partitioning of the tetrahedral intermediate to either products or reactants will be shifted towards the former.

A question arises upon examination of these bifunctional hydrolysis catalysts of Motomura and Haake: are simple organic diacids capable of similar rate accelerations when present in their monoanionic form? Monoanions of adipic and succinic acid were investigated both with molecular modeling and experimentally in the hydrolysis of *p*NPA. Interestingly, the monoanion of simple diacid succinic acid appeared by models[34] not to have a stable conformation with hydrogen bonding within the tetrahedral intermediate between the alcoholic leaving group oxygen and the protonated carboxylic acid of the monoanionic catalyst. However, the intermediate formed with adipic acid monoanion, having a tether between the acid and base two carbons longer than succinic acid, as the nucleophile, does have an energy minimum with the previously described hydrogen bond. However, a stochastic dynamics simulation indicates that the entropy loss upon forming this hydrogen bonded structure is too great to overcome with any resulting enthalpically favorable interactions. These results were borne out experimentally where neither of these diacid monoanions exhibited remarkable reactivity based on their pK_a values.[35]

8.5 Glycosidic Bond Hydrolysis

Catalysis of acetal hydrolysis was intently studied for many years. During much of this time, the hydrolysis of acetal linkages was found to be catalyzed exclusively via specific acid catalysis where the suspected mechanism consisted of a pre-equilibrated protonation of the acetal oxygen followed by rate-determining expulsion of an alcoholic leaving group and simultaneous formation of a carbenium ion intermediate.[36] Specific acid catalysis is only dependent upon the hydronium ion concentration and not on the concentration of any buffer acids present, as in the case of general acid catalysis. The interest in observing general acid catalysis of acetal hydrolysis stems from the hypotheses regarding glycosidase enzyme action and the notion that weak carboxylic acids are capable of producing rate accelerations of up to 10^{17} times over spontaneous aqueous hydrolysis.[2,3]

Fife discovered a substrate capable of undergoing buffer acid catalyzed hydrolysis,[37] and additional examples soon followed.[38] In order to facilitate general acid catalysis, the substrates had to be designed to have very good leaving groups so that protonation of the acetal oxygen became rate limiting, as in the case of *p*-nitrophenoxy tetrahydropyran.[39] Otherwise, specific acid catalysis with rate-limiting C–O bond cleavage would occur.

Perhaps the most compelling of these examples, however, comes from the intramolecular hydrolysis of Capon.[40–42] This system is interesting because it performs remarkably well in water even in the absence of a hydrophobic binding pocket. The key to the hydrolytic success of these molecules appears to be the precise positioning of reactive functionalities.

8.5.1 Intramolecularity

In intramolecular systems, the catalyst and the substrate are part of the same molecule, which facilitates interactions between them. Subsequently, intramolecular reactions of many types generally exhibit large rate enhancements over their intermolecular counterparts.[9] While several factors can in principle contribute to this result, such as steric or electronic effects, the most relevant for our purposes is the high effective local concentration. This stems from the remediation in the intramolecular case of the negative entropy associated with bringing a reactant and catalyst together during an otherwise intermolecular reaction. It is important to note that with an intermolecular system comes some complexity that must be addressed before arriving at any particular interpretation of the relevant kinetics. Steric and electronic effects, for instance, have been previously misinterpreted as bifunctional catalysis[43–46] in intramolecular ester hydrolysis reactions. Even with this added complexity, however, there are several pertinent examples that are relevant to glycosidic bond hydrolysis and biofuels.

Capon synthesized *o*-carboxyphenyl β-D-glucoside (Figure 8.7) which underwent hydrolysis 13 000 times faster than the electronically similar *p*-carboxyphenyl isomer at pH 4.55.[40] In fact, the hydrolysis of this *p*-substituted substrate was so slow under mild conditions of pH 4.55 that its hydrolysis rate had to be extrapolated from experiments at more acidic pH. The salicylate

Figure 8.7   *o*-Carboxyphenyl β-D-glucoside substrate capable of intramolecular acid catalyzed hydrolysis (adapted from ref. 40).

Figure 8.8 *o*-Carboxyphenyl 2-acetamido-2-deoxy-D-glucose substrate capable of undergoing intramolecular acid–base bifunctional catalysis (adapted from ref. 47).

leaving group possesses an intramolecular hydrogen bond, indicating that the carboxylic acid group of *o*-carboxyphenyl β-D-glucoside is in the correct position to donate a proton to the glycosidic oxygen and facilitate C–O bond breaking. In a similar study, Capon was also able to observe intramolecular bifunctional catalysis of glycosidic bond hydrolysis (Figure 8.8) while mimicking the action of β-*N*-acetylhexosaminidases.[47] In this case, the carboxylic acid still serves to protonate the alcoholic leaving group, and now the amide oxygen is in a position to act as a nucleophile towards the carbon in the 1 position of the hexose unit. The hydrolysis of this compound is 20-fold faster than that of the similar substrate without an acetylamido group, indicating bifunctional catalysis with both the acidic and basic moieties contributing.

While induced intramolecularity certainly produces interesting results and further confirms the importance of precise positioning when designing a bioinspired catalyst, the applications to practical catalytic systems are limited. In general it is not feasible to design a substrate to have a 'built-in' catalyst. A challenge for the design of synthetic catalysts is to find a way to implement the features of what makes the intramolecular system so successful, into a catalyst capable of acting intermolecularly. Capon's compounds are essentially an extreme version of the binding implemented in the previously described bifunctional catalysts, and can be viewed as systems where the equilibrium is shifted in such a way as to have binding in a chemisorbed state as in Figure 8.1c. It is difficult to design a catalyst, particularly a multifunctional catalyst, where all of the relevant functional groups are positioned precisely correctly.

However, this issue can be circumvented by implementing a catalyst where the reactive groups are present at a continuum of close distances relative to one another. By having a number of possible catalytic configurations present, there is a high likelihood that at least one will make for an effective catalyst. This argues for synthesis of catalytic sites along a polymer strand or on a two-dimensional surface. This is a similar principle to what was observed by Hine during the (poly)ethylenimine (PEI) catalyzed dedeuteration of Me_2CDCHO in water.[48,49] When an iminium ion was formed between the deuterated iso-butyraldehyde and a protonated amine of the PEI, adjacent amines were found to effectively catalyze its dedeuteration. It was clear the catalytically active

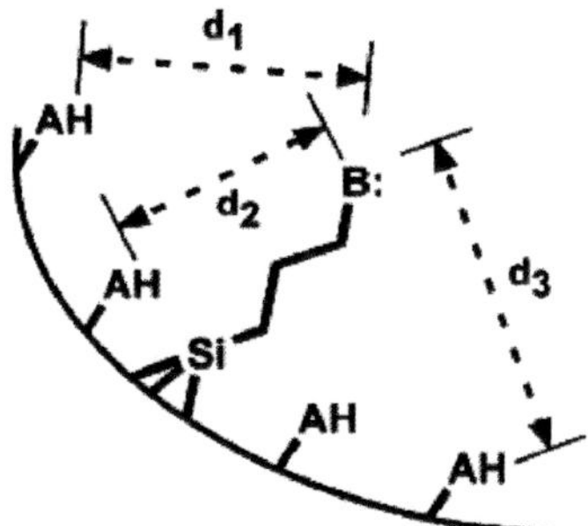

Figure 8.9 The silica surface as a macroscopic ligand providing a range of acid–base distances for bifunctional cooperativity (reproduced with permission from ref. 48).

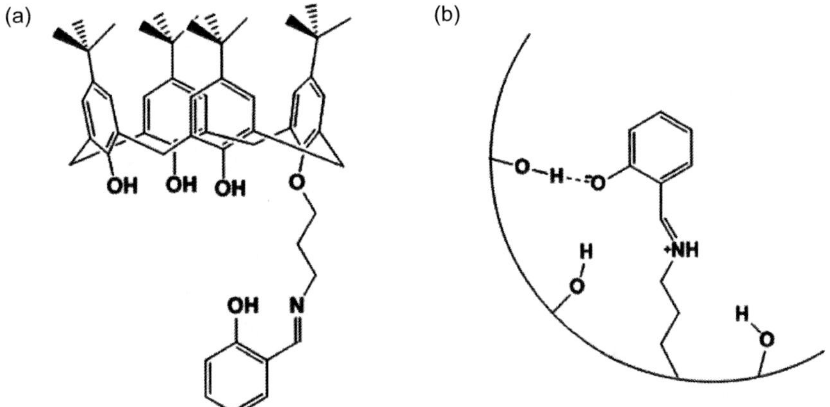

Figure 8.10 Calixarene Schiff base in phenolic form (a) and Schiff base of aminopropyl functionalized silica in zwitterionic form facilitated by the presence of acidic surface silanols (b) (adapted from ref. 48).

amines were intrapolymer because the rate reached a maximum with increasing PEI concentration. Hine called this the 'shotgun approach' because the polymer provided a continuum of distances between imines and amines so that there was a high probability that at least one amine would be in the correct position, despite the stringent spatial requirements of bifunctional catalysis.[50] This approach was implemented in section 8.6.2 with functionalized inorganic oxide surfaces, which also provide a continuum of distances between reactive functional groups.

To further highlight this point, take for instance the organocatalysts studied by Bass and Katz. They found that a homogeneous monoaminated calixarene (Figure 8.10a) is unable to support the zwitterionic tautomer of the Schiff base formed upon reaction of the amine with salicylaldehyde in non-polar solvent.[51] While the calixarene still possesses an acidic phenolic proton in proximity to the

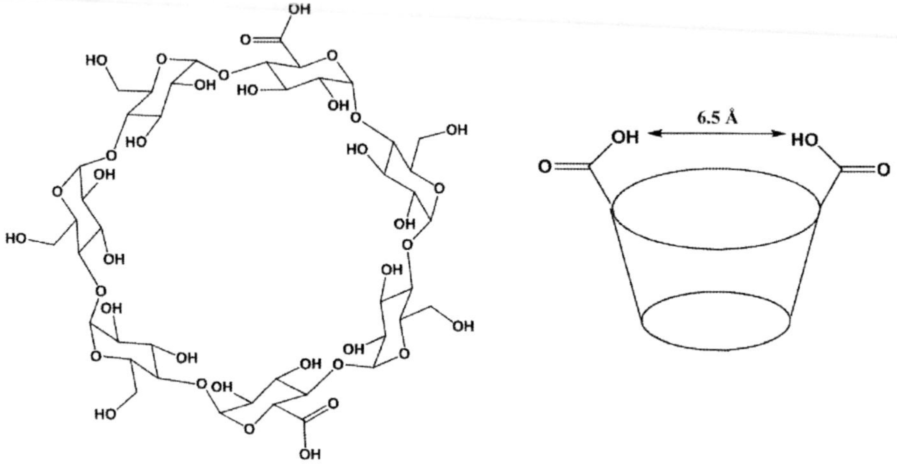

Figure 8.11 Dicarboxylic acid functionalized β-cyclodextrin glycosidase mimic (adapted from ref. 51).

imine, the positioning is not precise enough to effectively interact with the imine and stabilize the zwitterionic tautomer. This is despite the similarity in pK_a of this acidic calixarene proton[52] relative to the pK_a of silanol on silica.[53] This is in stark contrast to the situation on silica containing active silanols (Figure 8.10b) where, when functionalized with a low enough density of aminopropyl groups to maintain a high coverage of acidic surface silanols, the zwitterionic tautomer of the Schiff base of salicylaldehyde predominates over the phenolic form. The zwitterionic form is facilitated and stabilized by the continuum of distances between acidic and basic sites on the surface of aminopropyl-functionalized silica. Catalytic applications of these materials in aldol reactions are discussed in greater detail in Section 8.6.

8.5.2 Cyclodextrins

The platform of functionalized cyclodextrins alluded to earlier has been recently applied to the synthesis of glycosidase enzyme mimics. Bols *et al.* synthesized both α- and β-cyclodextrins bearing several combinations of functionalities directed towards the binding cavity.[54–56] The first attempt was the most faithful reproduction of the active site of a glycosidase and consisted of cyclodextrins functionalized with two carboxylic acids (Figure 8.11).[54] The distances between the acids were 5 and 6.5 Å in the α- and β-cyclodextrins, respectively, which are on par with the estimated distances between the reactive Glu and Asp residues in a retaining glycosidase. The activity of these catalysts was assessed by the hydrolysis of *p*-nitrophenyl β-glucopyranoside, an activated analog to hexose glycosidase substrates. However, aside from Michaelis–Menten kinetics, the reactivity of these catalysts bore little resemblance to the high activity of a glycosidase enzyme.[57] In another attempt to produce a hydrolytic cyclodextrin catalyst, Ortega-Caballero *et al.* synthesized α- and β-cyclodextrins functionalized with cyanohydrin groups positioned so that the acidic –OH functionality was oriented towards the binding cavity.[55] These catalysts are suspected to engage in unactivated binding with the glucoside substrate along with protonating the *p*-nitrophenol leaving group, facilitating up to a 2000-fold rate enhancement of glycoside hydrolysis when compared to the background reaction in water. Finally, in a logical extension of the previously described catalysts, Ortega-Caballero and Bols attempted to synthesize multifunctional cyclodextrins with both cyanohydrins and carboxylate groups in an attempt to recreate the acid–base chemistry of glycosidases.[56] Unfortunately, however, these compounds proved to either be less effective than their monofunctional ancestors or too unstable to obtain reliable catalytic data.

8.5.3 Catalytic Antibodies

Catalytic antibodies[58] incorporate binding elements desirable for catalyzing a chemical reaction akin to the lock and key principle of Emil Fischer.[59] They are synthesized by a multistep process. First, certain features of a substrate or transition state (Figure 8.12a) expected to be important in an enzyme-catalyzed reaction are identified and recreated in a hapten (Figure 8.12b). The hapten is an organic molecule which can elicit an immune system response when coupled with a carrier protein and introduced to an organism. Therefore, when features that are expected to be meaningful in an enzymatic transition state are replicated in a hapten, the immune system response to that antigen can be steered to make antibodies that engage in either activated or unactivated binding with substrates similar to the hapten used. Conveniently, since many antibodies are made in response to the stimulus, there is a high probability that at least one of these will have the precise positioning required to be an effective catalyst.

(a)

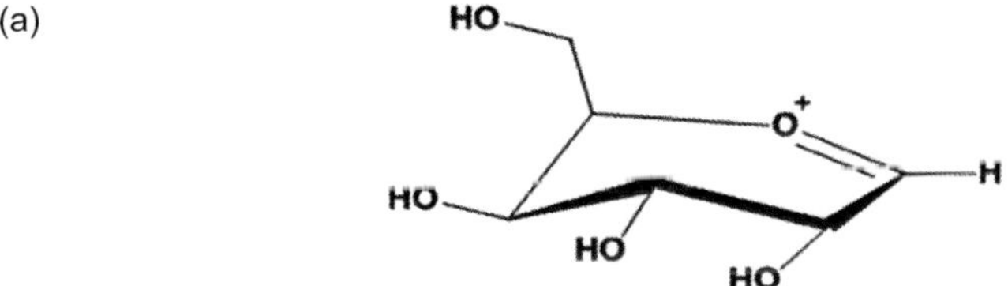

(b)

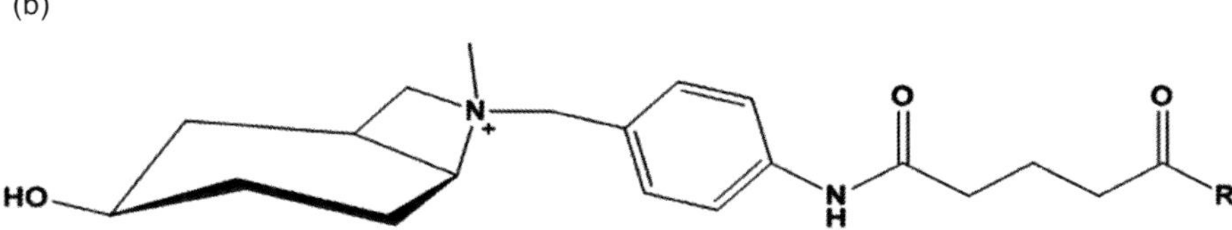

Figure 8.12 Carbenium ion intermediate (a) formed during rate-limiting C–O bond cleavage and hapten (b) designed to replicate half-chair conformation and positive charge of (a). R = *N*-hydroxysuccinimide (adapted from ref. 59).

Therefore, the combinatorial nature of catalytic antibodies obviates some of the difficulties associated with designing and synthesizing individual bioinspired catalysts.

Several features of suspected importance to glycosidase enzyme action have been targeted and replicated in haptens.[60–64] The first is the distortion of a reacting hexose unit into a half chair geometry. As alluded to earlier, some of the catalytic success of hydrolytic enzymes is suspected to stem from the enzyme's ability to distort the substrate into a more reactive conformation. The half chair is of higher energy and closer in geometry to the carbenium ion formed during rate-limiting C–O bond breaking. This distortion should therefore lower the activation energy of alcoholic leaving group expulsion. The thought is that a hapten mimicking such a conformation will induce the immune system to synthesize proteins capable of accommodating this feature of the proposed transition state. Another feature of hydrolytic enzymes suspected to be of catalytic significance is a negatively charged Asp residue at the active site. This moiety is expected to act either as a basic catalyst in a bifunctional mechanism with the acidic action of the nearby Glu residue, or through electrostatic stabilization of developing positive charge in the transition state. To introduce this type of feature into a catalytic antibody, haptens have been designed with precisely positioned quaternary ammonium cations to replicate the presence of positive charge in the protonated glycosidic substrate. With these design principles in place, several antibodies capable of hydrolyzing activated glycosidic bonds have been generated and isolated. Antibody 14D9, among the most active of those isolated from these studies with a maximum rate enhancement of 70 times over the background reaction,[54] was later identified as having ionizable groups at the active site. Interestingly, the absence of a kinetic isotope effect for 14D9-catalyzed glycosidic bond hydrolysis indicates that these functionalities were not acting as general acids, but could perhaps be acting as basic catalysts or as stabilizers to the developing positive charge resulting from a specific acid catalyzed reaction pathway.[63]

In principle, by creating a library of haptens, one could analyze the resulting antibodies for activity and learn more about the influence of various substrate–catalyst interactions on reaction rates. Other advantages to this combinatorial approach are avoiding the necessity of precisely synthesizing a catalyst that must meet very stringent organizational requirements and negating interference from water by creating, like an enzyme, a non-polar binding pocket for substrate molecules. Conversely, the pathways to obtain the catalysts are relatively cumbersome and their success is ultimately still limited by the synthetic techniques employed during hapten fabrication. Additionally, their structures can be as complex as the enzymes that inspired their creation, making structure–function relationships a practical difficulty.

8.5.4 Combinatorial Polymer Catalysts

Similar principles to those guiding the design of haptens and the resulting catalytic antibodies can also be applied to entirely synthetic systems. This has been motivation behind synthesis of molecularly imprinted polymers. These procedures have been reviewed elsewhere,[65] but in general they involve the formation of a polymeric catalyst around a transition state analog (TSA) which is then removed, leaving behind catalytically active binding sites with specificity for the transformation mimicked by the TSA. Like catalytic antibodies, this approach involves the 'reverse engineering' of catalysts by starting with a template representative of the suspected transition state, and having catalysts form in response to the TSA or hapten. The heterogeneity of sites inherent to molecularly imprinted polymers is useful because it is likely that at least some of the active sites will be in the correct configuration so as to be effective; however, without any single well-defined active site, the specific determination of mechanistic details is a significant challenge.

While there is no known example of a glycosidase mimic produced by molecularly imprinted polymers, a similar method, dynamic combinatorial library selection (DCL), has been used to synthesize glycosidase mimics capable of catalyzing acetal hydrolysis. DCL is based on shifting the equilibrium of molecules formed from a mixture of monomers via reversible bonds, typically disulfide, towards those stabilized by interaction with a TSA. Once equilibrated, the reactions are quenched and the resulting catalysts are isolated. With the use of a TSA for the formation of a carbenium ion from a model acetal, Vial *et al.* were able to isolate a macrocyclic catalyst capable of enhancing the rate of the hydrolysis of a model acetal to over twice the uncatalyzed rate, while each of the monomers alone showed no activity.[66] The catalyst was active in water, and some of this success can be attributed to hydrophobic interactions between the aryl acetal and the macrocyclic catalyst interior.

8.6 Aldol Condensations

Aldol reactions represent an important class of reactions occurring during the aqueous phase reforming of highly oxygenated biomass-derived feedstocks.[67] These C–C bond-forming reactions create large molecules which can be further processed into liquid hydrocarbon fuels.

8.6.1 Homogeneous Organocatalysis

In comparison to controversial catalytic systems where the presence of bifunctional catalysis has been harshly doubted in the past, to the point of shedding the whole field in an unbalanced negative light,[15,46] the body of evidence in support of bifunctional catalysis of aldol condensations is overwhelming. Take for instance the organocatalysis of proline, which mimics the behavior of enamine-based aldolases. List *et al.* studied proline and several analogs to elucidate the requirements for the efficient catalysis of the aldol reaction between acetone and *p*-nitrobenzaldehyde.[68] When either the pyrrolidine amine or carboxylate were removed, the yield after 24 h decreased from 68% to below 10% for most of the catalysts studied. Clearly both

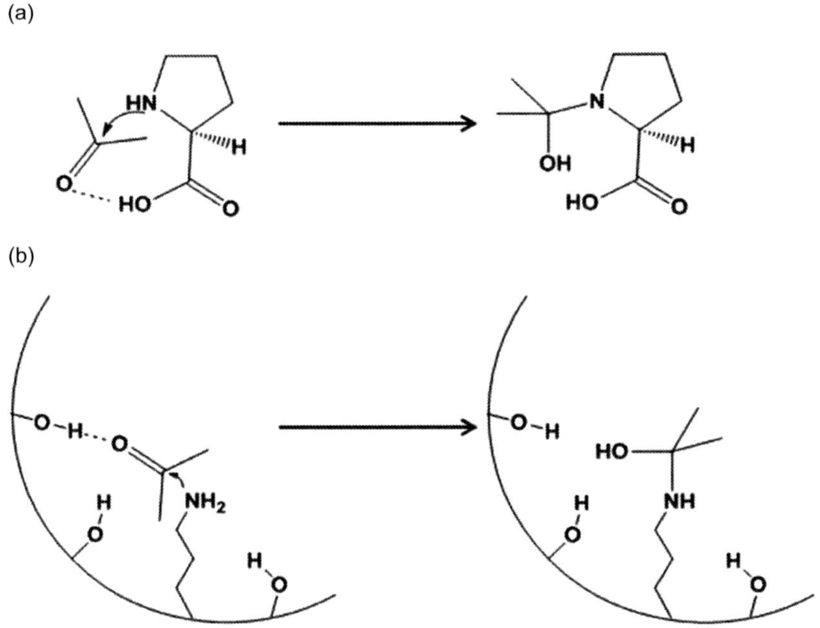

(a)

(b)

Scheme 8.1   Outline of nucleophilic amine addition, one of several steps suspected to be subject to bifunctional catalysis during aldol reactions, during homogeneous proline catalysis (a) and heterogeneous aminosilica catalysis (b) (adapted from refs. 68 and 48).

functional groups are responsible for the proficient catalysis of the aldol reaction by proline, with the carboxylate group participating in activated binding as depicted in Figure 8.1b. It is hypothesized that the carboxylate group can assist in basic catalysis by the amine via activation of the carbonyl substrate to nucleophilic attack (Scheme 8.1a), deprotonation of the iminium ion formed upon dehydration of the product of nucleophilic addition, activation of the nitroaldehyde to nucleophilic attack by the unsaturated proline compound, and/or facilitation of the hydrolysis of the iminium aldol intermediate to form the final aldol product. While proline itself is not an effective catalyst in the presence of a large excess of water, there has been recent success[69] with bifunctional catalysis of the aldol reaction in water by appending proline derivatives with hydrophobic groups to engage in unactivated binding with non-polar substrate functionalities.

8.6.2   Heterogeneous Amine-functionalized Silica

The continuum of acidity on the silica surface greatly increases the odds over homogeneous counterparts that at least one of the acids will be in the proper orientation to act cooperatively with a basic surface functionality. This relieves the catalyst designer of the burden of precisely synthesizing a catalyst with exactly the correct structure to facilitate bifunctional cooperativity. As was seen, aldol condensation reactions are well-suited to bifunctional catalysis, and bifunctional systems satisfy the desire to operate in a regime where the reactions can proceed at neutral conditions of temperature, pressure, and pH.

A similar environment to that created by the homogeneous proline catalyst is reproduced by the previously described aminated silica materials (Scheme 8.1b).[51] Catalysis of the nitroaldol condensation reaction between *p*-nitrobenzaldehyde and nitromethane results in a dramatic reversal in selectivity when compared to monofunctional catalysis. The acidity of the silanol-rich surface facilitates the formation of an iminium ion intermediate (see above), which leads to unsaturated products as the water is irreversibly lost during imine formation. In similar heterogeneous amino catalysts (derived from the same parent material), when silanols native to silica are removed via capping, iminium ion formation is no longer favorable and the reaction proceeds through an ion pair mechanism involving charge-separated intermediates, to produce alcohol products. Further evidence of the bifunctional environment present in these materials is provided by the related Knoevenagel condensation between *m*-nitrobenzaldehyde and malononitrile. Materials with surface silanols exhibit an order of magnitude rate enhancement over those capped with either polar or non-polar functionalities. Interestingly, the bifunctional materials maintain their activity in the presence of water. This means formation of the iminium ion intermediate is uninhibited by water in these systems.

Zeidan and Davis probed the effect of functionalizing silica with moieties more acidic than the native silanols to interact with the surface amines.[7,70] In addition to amines, three different acid groups, sulfonic, phosphoric, and carboxylic, with pK_a values of approximately –2, 3, and 5, respectively, were incorporated into the silica materials with the aim of improving over the activated binding exhibited by silanols in the aldol condensation between *p*-nitrobenzaldehyde and acetone. For reference, the average pK_a of a native silica silanol is 7. In solution, these functionalities completely neutralized one another and resulted in catalytically inactive salts. When both the acid and base were immobilized on SBA-15, however, the conversions were 62%, 78%, and 99%, for the sulfonic, phosphoric, and carboxylic acid groups, respectively, after 20 hours. This type of activity trend favoring weaker acids to be paired with amines as bifunctional catalysts has been observed in related systems as well.[71] Control experiments demonstrate significantly reduced catalysis by materials functionalized with either amines, acids, or a physical mixture of these monofunctional materials (Table 8.1).[7]

While all of the materials with functionalities more acidic than silanols exhibited an enhanced reactivity, clearly the carboxylic acid material was the most highly effective partner for the amine in the bifunctional system described above. This is interesting because it is also the most weakly acidic of the three functionalities tested and is the choice of acid group used by biological catalysts. It seems as if the equilibrium between free acids and bases and the neutralized ion pair is largely dependent upon the strength of the acids used. Sulfonic acid, being the strongest acid studied, shifts the equilibrium towards the neutralized ion pair as compared to the carboxylic acid, which, due to its

Table 8.1 Catalytic data for bifunctional catalysis of an aldol condensation with acids of varying strength paired with primary amines on silica. (Reproduced with permission from ref. 8.)

Entry	Catalyst (10 mol%)	% A	% B	Total conversion
1	SBA-15—(CH₂)₃NH₂ and SBA-15—CH₂CH₂—C₆H₄—SO₃H	45	17	62
2	SBA-15—(CH₂)₃NH₂ and SBA-15—CH₂CH₂—P(=O)(OH)OH	62	16	78
3	SBA-15—(CH₂)₃NH₂ and SBA-15—CH₂CH₂—C(=O)OH	75	24	99
4	SBA-15—CH₂CH₂—P(=O)(OH)OH	0	0	0
5	SBA-15—CH₂CH₂—C(=O)OH	0	0	0
6	SBA-15—CH₂CH₂—C₆H₄—SO₃H	8	8	16
7	SBA-15—(CH₂)₃NH₂	25	8	33

weaker acidity will be more apt to coexist with the amine in the catalytically active free acid form. This equilibrium is also greatly influenced by the solvent. In water, the conversion for a given time was reduced to less than half the value obtained in non-polar aprotic solvents. This decrease in activity is attributable to both the stabilization of the neutralized ion pair by water and by the efficient proton conductivity of aqueous solution. Therefore, the effectiveness of these catalysts in non-polar solvents is analogous to the success of these same functional groups in the hydrophobic enzymatic binding pockets. Both here and in the enzymes, the weakly acidic species exhibit an enhanced reactivity compared to more acidic species, which have a higher propensity towards neutralization and also tend to create environments with undesirably harsh pH conditions.

The materials of Katz *et al.* rely on a combinatorial approach in which the surface enforces many possible acid–base distances, only a few of which are the "right" ones for the particular catalysis at-hand (*vide supra*). Such a mechanism of action is also likely followed in other related materials such as those described by Davis *et al.* (*e.g.*, in control catalysts consisting of aminopropyl groups tethered off of the surface of silica).[7] An interesting observation is that the successful homogeneous systems described above, consisting of the cyclo-dextrin system by Rao *et al.*[19] and the ascorbic acid system of Shaskus and Haake[33] both consist of a nucleophile active site surrounded multiple OH groups, which may act to enforce the right distance for leaving group activation *via* hydrogen bonding. A hypothesis based on this observation is that all of these systems above benefit from the same combinatorial approach of having a high local density of OH groups in the vicinity of the nucleophilic center.

8.7 Ketonization

Catalysis of ketonization, another relevant reaction during the reforming of oxygenated biomass, benefits from bifunctional catalysts in a different, but equally as effective, fashion. Ketonization is known to be catalyzed by bifunctional alkali-doped zirconia and titania.[72] It is believed that a two-step process occurs where two carboxylic acid groups are condensed to form a ketocarboxylic acid intermediate, which subsequently undergoes CO_2 elimination. It is hypothesized that the former reaction is base catalyzed while the latter is acid catalyzed. Therefore, a bifunctional catalyst capable of performing both of these successive steps is extremely convenient and illustrates how bifunctional catalysts are capable of cascade reaction systems, where the acid and base can facilitate separate reactions en route to the final product. Recently, Gaertner and co-workers studied the ketonization of carboxylic acid feedstocks similar to those found during the reforming of biomass to fuels.[73] In these studies, ceria-doped zirconia proved to be an efficient catalyst for the ketonization and esterification of hexanoic acid in alcoholic solution, catalyzing both the direct ketonization of carboxylic acids, and at higher temperatures, the subsequent ketonization of any esters formed. Since these are cascade reactions (i.e. reactions in series) and the catalyst is not bifunctional in the activated binding sense, competitive solvation of reactants by water is not an issue here, and any inhibition from water is due to its strong adsorption on ceria–zirconia base sites which hinders the ketonization reaction.

8.8 Dehydration

As alluded to earlier, the upgrading of highly oxygenated biological feedstocks requires processes for efficient deoxygenation. Zirconia, a bifunctional catalyst with a rich history of effectively performing many transformations,[75,76] has been identified as being an efficient and selective heterogeneous catalyst for the dehydration of several alcohols to α-olefins.[76,77] In fact, Tanabe postulates that any reaction subject to either acid or base catalysis is a candidate for bifunctional catalysis if the properties of the catalyst are suitable.[78] Additionally, the weak acidity and basicity of zirconia give these catalysts characteristically better stability than more acidic or basic materials. To illustrate the effectiveness of these catalysts, take for instance the dehydration of 1-cyclohexylethanol to vinylcyclohexane (Figure 8.13). Over ZrO_2, this reaction proceeds to higher yields than when catalyzed by other more acidic or basic catalysts. This is suspected to be attributable to bifunctional catalysis because the activity is diminished by either n-butylamine or carbon dioxide adsorption.[74] More recent studies of dehydration catalysis involving tunable Keggin-type polyoxometalate (POM) as acid catalysts have demonstrated compensating mechanisms to explain the observed catalytic proficiency of weak acids in dehydration reactions, when compared with other reactions such as isomerizations.[79] Iglesia *et al.* elegantly demonstrate that this can result due to the ability of the conjugate base (anion) of the weak acid of the POM to participate directly in stabilizing charge separation at the transition state, due to the local nature of charge build-up in the dehydration transition state, which otherwise compensates for a weaker POM proton donating ability. This type of mechanism involving both the acidic hydrogen as well as the conjugate base within the mechanism can indeed be viewed as bifunctional or cooperative catalysis. However, in related systems, Iglesia *et al.* has shown that the formation of catalytically inactive dimers involving POMs is disfavored for weaker acids, and suggests this to be an additional compensating mechanism that tends to decrease the sensitivity of POM acid strength on dehydration catalysis.[79,80] This scenario involves two different mechanisms, one of which represents catalyst deactivation, and does not require bifunctional catalysis between acid and conjugate base. The examples above of understanding dehydration catalysis point to the complexity of possible mechanisms and caution against ascribing dehydration by weak acids as pointing necessarily to a bifunctional mechanism.

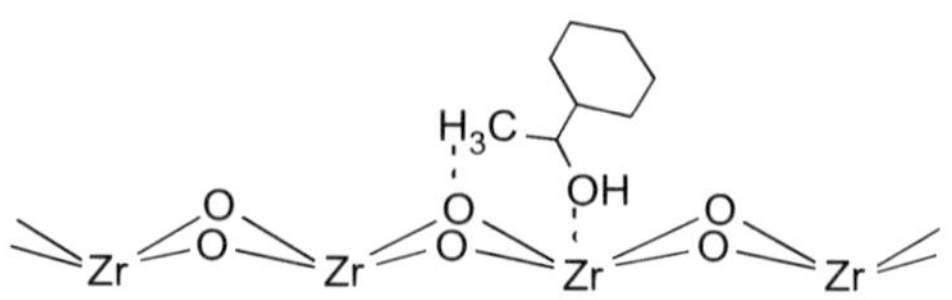

Figure 8.13   Schematic representing similarity of distances between acidic Zr^{4+} and basic O^{2-} in 400 °C pretreated zirconia and between basic hydroxyl and acidic methyl hydrogens of 1-cyclohexylethanol (adapted from ref. 73).

8.9 Lignin Depolymerization

Kidder *et al.* recently illustrated the effect of an acidic oxide surface on another reaction relevant to biofuel feedstock processing, lignin depolymerization.[81,82] Model lignin compounds, phenalkyl phenyl ethers, were covalently bound to mesoporous silicas and the influence of the silica surface on pyrolysis of the ethers was assessed. Interestingly, it was discovered that the grafting density had a profound effect on the product selectivity. When loaded on SBA-15 at 0.74 phenalkyl phenyl ether molecules per nm,^{2} the ratio of alcoholic products to aldehyde products was 14.2. However, when the density was lowered to 0.28, the ratio increased to 25.7.[81] It was postulated that increasing the grafted surface density of ether on silica removed surface silanols and disrupted ligand–surface interactions. This change in spatial configuration is expected to alleviate steric hindrance in the production of the β product. It is also important to note that molecular modeling studies indicate very prominent hydrogen bonding between the ether oxygen and surface silanols. Once again the silica surface provides an acid-rich environment with high enough local silanol densities to greatly influence surface chemistry. The continuum of distances present between phenyl ethers and surface silanols gives this system a very high probability of having at least one productive interaction. Furthermore, previous studies showed that in addition to impacting selectivity, the rate of pyrolysis can be enhanced by the incorporation of ligands into mesoporous as opposed to nonporous silica, and even further impacted by the size of the pores.[82] The pyrolysis of 1,3-diphenylpropane (DPP) immobilized on mesoporous MCM-41 occurs 5-fold faster than that grafted onto nonporous cabosil at the same surface density. It is hypothesized that mesoporous silica allows the DPP to be in closer proximity to one another which facilitates radical hydrogen transfer.

In addition to acting alone during lignin depolymerization, preliminary results suggest that benzyl ether hydrolysis is facilitated by the action of Brønsted acidic silanols in concert with the Lewis acidity of titanium.[83] When a homooxacalix[3]arene is refluxed in toluene with either silica or titanium isopropoxide alone, no intracalixarene ether cleavage occurs. However, when refluxed in the presence of both titanium and silica, the ether linkage is broken as evidenced by the formation of an aldehyde seen by proton NMR and FT-IR. This shows that cooperativity between Lewis acidic Ti centers and Brønsted acidic silanols is enough to cause intramolecular benzyl ether cleavage, whereas each type of acid site functioning on its own is not. Here, the ether cleavage reaction is facilitated by the range of distances between silanols and Ti centers that is offered by the silica surface, which facilitates cooperativity between the two types of acid sites.

If the aldehyde generated in the paragraph above can then be decarbonylated, this procedure represents a critical development in the synthesis of energetic fuels via deoxygenation through carbon loss as CO or CO_2. Biological systems tend to deoxygenate compounds through this pathway; however, most synthetic processes involve deoxygenation by removal of water, which negatively impacts the hydrogen economy of the process. With the large variety of configurations of silanols on the silica surface, there is a much higher likelihood that the homooxacalix[3]arene will engage in a productive interaction with both the titanium and the silica than if there were only one or few possibilities as would be the case with a homogeneous Brønsted acid catalyst.

8.10   Conclusions and Future Directions

The efficiency with which nature operates makes learning from enzymes a critical approach for understanding and enhancing synthetic catalysts. Enzymes are unparalleled in their abilities to catalyze reactions with the utmost selectivity and proficiency. Enzymes relevant to biofuels production, glycosidases and lipases for instance, contain within their design the blueprints for as of yet unrealized synthetic catalysts. Any information learned about enzymatic transformations can be, and should be, applied to the design of new catalytic materials.

Bifunctional catalysis has established itself to be one of the more intriguing aspects of enzymatic catalysis. Many chemical systems have been developed to reproduce enzymatic binding, either unactivated or activated, to go along with a catalytic action. The interplay between multiple functional groups presents an attractive approach to many catalytic challenges because multiple groups working in concert allows reactions to proceed with as good or better efficiency then when catalyzed in much harsher conditions of temperature, pressure, and pH. Because bifunctional catalysts operate by mechanisms that do not obey activity correlations set forth for their monofunctional counterparts, it is important to note that predictions for what is and is not possible for bifunctional systems should not be based on trends established for monofunctional catalysts.[33,84] Another intriguing aspect of enzymatic catalysis is the hydrophobic nature of many natural binding sites. By excluding or selectively including water, and forming covalent intermediates where necessary, enzymes exert absolute control over all binding interactions and are capable of efficiently performing even the most selective of transformations.

Several different approaches, all with their own advantages and drawbacks, have been implemented in the design of bioinspired catalysts. Perhaps the most common approach is that of rational catalyst design where a molecule is synthesized so as to have similar functionality as the target enzyme. Cyclodextrins typify this type of design scheme. With a hydrophobic binding pocket, cyclodextrins are able to exclude water from the active site and retain bifunctional activity in water, one of the main challenges facing bifunctional catalysts for biofuel applications.

Hydrophobicity is not a requirement of bifunctional catalysts, however, as was seen in the aqueous intramolecular β-glycosidic bond hydrolysis of Capon,[40] as well as the intermolecular transesterification of Hamilton,[25] the acyl transfer of Motomura,[27] and the ester hydrolysis of Haake,[33] all occurring in either water or a polar solvent (acetonitrile). Clearly the proper positioning of reactive functionalities is the real key to successful bifunctional catalyst synthesis. This is not an easy task, however, and while these catalysts are interesting in the sense that their structures are in general very well defined, making structure–function relationships possible, it is very difficult to get the precise positioning of all relevant functionalities correct. This is clearly illustrated by Hine's observation that by changing the distance between acid and base during the catalysis of aqueous imine formation by one methyl group, the rate drops by an order of magnitude.[84]

This problem of positioning can be aided by using a combinatorial system, such as catalytic antibodies or molecularly imprinted polymers, where many different catalysts are made, from which the most effective can be purified and analyzed. While this certainly relieves much of the synthetic burden, the resulting materials are typically not very well-defined, and the process to obtaining them is still quite cumbersome.

An alternative combinatorial approach to circumvent having to precisely recreate the complexity of an enzyme is to turn to systems where there exists a continuum of distances between reactive functionalities. This way, it is quite likely that at least one of the necessary catalytic groups will be in the proper position to react in the desired fashion. Hine demonstrated this effect with polymers during the dedeuteration of acetone,[48] and Bass and Katz were able to demonstrate a similar phenomenon with two-dimensional surfaces employing primary amines and silanols.[51] The advantage of the latter being a two-dimensional network of sites that are more closely spaced given the shorter Si–O–Si repeat distance compared with an organic polymer.

A crucial remaining challenge in this area of biofuels catalysis is how to convert the intramolecular system of Capon[40] described above to an intermolecular catalysis system which hydrolyzes β-glycosidic bonds in cellulose directly. Future work should focus on developing systems that rely upon a continuum of distances between reactive functionalities, seeing as they, by sheer strength in numbers, have a much higher probability for success than those coming from a method involving designing, synthesizing, and evaluating catalysts individually. It is a practical impossibility to recreate one-by-one the nearly endless number of active site configurations present in one functionalized oxide material. Now with quite a bit of history and progress behind it, the area of bioinspired catalysis truly represents a crucial piece of the future of biofuels processing.

References

1. A. Radzicka and R. Wolfenden, *Science*, 1995, **267**, 90.
2. C. S. Rye and S. G. Withers, *Curr. Opin. Chem. Biol.*, 2000, **5**, 573.
3. R. Wolfenden, X. Lu and G. Young, *J. Am. Chem. Soc.*, 1998, **120**, 6814.
4. C. C. F. Blake, D. F. Koenig, G. A. Mair, A. C. T. North, D. C. Phillips and V. R. Sarma, *Nature*, 1965, **206**, 757.
5. L. N. Johnson and D. C. Phillips, *Nature*, 1965, **206**, 761.
6. M. A. Raftery and T. Rand-Meir, *Biochemistry*, 1968, **7**, 3281.

7. R. K. Zeidan and M. E. Davis, *J. Catal.*, 2007, **247**, 379.
8. E. L. Margelefsky, R. K. Zeidan and M. E. Davis, *Chem. Soc. Rev.*, 2008, **37**, 1118.
9. M. I. Page and W. P. Jencks, *Proc. Natl. Acad. Sci.*, 1971, **68**, 1678.
10. R. Henderson, *J. Mol. Biol.*, 1970, **54**, 341.
11. R. Henderson, C. S. Wright, G. P. Hess and D. M. Blow, *Cold Spring Harbor Symp. Quant. Biol.*, 1971, **36**, 63.
12. B. D. Sykes, S. L. Patt and D. Dolphin, *Cold Spring Harbor Symp. Quant. Biol.*, 1971, **36**, 29.
13. H. J. Schneider, *Angew. Chem., Int. Ed. Engl.*, 1991, **103**, 1417.
14. R. Breslow, *Acc. Chem. Res.*, 1995, **28**, 146; plus references therein.
15. W. P. Jencks, *Catalysis in Chemistry and Enzymology*, McGraw-Hill, New York, 1969.
16. C. G. Swain and J. F. Brown, Jr., *J. Am. Chem. Soc.*, 1952, **74**, 2538.
17. J. M. Notestein, A. Katz and E. Iglesia, *Langmuir*, 2006, **22**, 4004.
18. F. Cramer and G. Mackensen, *Angew. Chem., Int. Ed. Engl.*, 1966, **5**, 601.
19. K. R. Rao, T. N. Srinivasan, N. Bhanumathi and P. B. Sattur, *J. Chem. Soc., Chem. Commun.*, 1990, **1**, 10.
20. F. R. Ma and M. A. Hanna, *Biores. Tech.*, 1999, **70**, 1.
21. C. H. Bamford and C. F. H. Tipper, *Comprehensive Chemical Kinetics*, Vol. 10, Elsevier, Amsterdam, 1972.
22. E. Minami and S. Saka, *Fuel*, 2006, **85**, 2479.
23. W. W. Cleland and M. M. Kreevoy, *Science*, 1994, **264**, 1887.
24. P. R. Schreiner, *Chem. Soc. Rev.*, 2003, **32**, 289.
25. P. Tecilla and A. D. Hamilton, *J. Chem. Soc., Chem. Commun.*, 1990, 1232.
26. V. Jubian, A. Veronese, R. P. Dixon and A. D. Hamilton, *Angew. Chem., Int. Ed. Engl.*, 1995, **34**, 1237.
27. T. Motomura, K. Inoue, K. Kobayashi and Y. Aoyama, *Tetrahedron Lett.*, 1991, **32**, 4757.
28. T. C. Bruice and G. L. Schmir, *J. Am. Chem. Soc.*, 1958, **80**, 148.
29. W. P. Jencks and M. Gilchrist, *J. Am. Chem. Soc.*, 1968, **90**, 2622.
30. M. L. Bender, *J. Am. Chem. Soc.*, 1951, **73**, 1626.
31. W. P. Jencks and J. Carriuolo, *J. Am. Chem. Soc.*, 1961, **83**, 1743.
32. (a) D. G. Oakenfull, T. Riley and V. Gold, *Chem. Commun.*, 1966. 385; (b) V. Gold, D. G. Oakenfull and T. Riley, *J. Chem. Soc. B*, 1968, 515.
33. J. Shaskus and P. Haake, *J. Org. Chem.*, 1984, **49**, 197.
34. T. J. Amundsen, K.A. Durkin and A. Katz, unpublished work. Maestro v. 9.0 and MacroModel v. 9.7 Schrodinger LLC. MC/SD simulations done with 1000 steps and OPLS_2005 forcefield.
35. T. J. Amundsen and A. Katz, unpublished work. Experiments performed with 100 µm *p*NPA and 10 µm succinic or adipic acid in methanol at 25 °C. Reaction progress monitored by *p*NPA disappearance observed *via* UV-vis absorption at 269 nm.
36. E. H. Cordes, *Progr. Phys. Org. Chem.*, 1967, **2**, 1.
37. T. H. Fife, *J. Am. Chem. Soc.*, 1967, **89**, 3228.
38. T. H. Fife, *Acc. Chem. Res.*, 1972, **5**, 264.

39. T. H. Fife and L. K. Jao, *J. Am. Chem. Soc.*, 1968, **90**, 4081.

40. B. Capon, *Tetrahedron Lett.*, 1963, **4**, 911.

41. B. Capon and M. C. Smith, *Chem. Commun.*, 1965, **7**, 523.

42. B. Capon, M. C. Smith, E. Anderson, R. H. Dahm and G. H. Sankey, *J. Chem. Soc. B 1969*, 1038.

43. H. Morawetz and I. Oreskes, *J. Am. Chem. Soc.*, 1958, **80**, 2591.

44. T. Higuchi, H. Takechi, I. H. Pitman and H. L. Fung, *J. Am. Chem. Soc.*, 1971, **93**, 539.

45. T. Maugh II and T. C. Bruice, *J. Am. Chem. Soc.*, 1971, **93**, 3237.

46. T. C. Bruice and I. Oka, *J. Am. Chem. Soc.*, 1974, **96**, 4500.

47. J. Hine, F. E. Rogers and R. E. Notari, *J. Am. Chem. Soc.*, 1968, **90**, 3279.

48. J. Hine, E. F. Glod, R. E. Notari, F. E. Rogers and F. C. Schmalstieg, *J. Am. Chem. Soc.*, 1973, **95**, 2537.

49. J. Hine, *Acc. Chem. Res.*, 1978, **11**, 1.

50. B. Capon, *Chem. Rev.*, 1969, **69**, 407.

51. J. D. Bass, A. Solovyov, A. J. Pascall and A. Katz, *J. Am. Chem. Soc.*, 2006, **128**, 3737.

52. K. Araki, K. Iwamoto, S. Shinkai and T. Matsuda, *Bull. Chem. Soc. Jpn.*, 1990, **63**, 3480.

53. M. L. Hair and W. Hertl, *J. Phys. Chem.*, 1970, **74**, 91.

54. C. Rousseau, N. Nielsen and M. Bols, *Tetrahedron Lett.*, 2004, **45**, 8709.

55. F. Ortega-Caballero, C. Rousseau, B. Christensen, T. E. Petersen and M. Bols, *J. Am. Chem. Soc.*, 2005, **127**, 3238.

56. F. Ortega-Caballero and M. Bols, *Can. J. Chem.*, 2006, **84**, 650.

57. C. Rousseau, F. Ortega-Caballero, L. U. Nordstrom, B. Christensen, T. E. Petersen and M. Bols, *Chem. Eur. J.*, 2005, **11**, 5094.

58. R. A. Lerner, S. J. Benkovic and P. G. Schultz, *Science*, 1991, **252**, 659.

59. E. Fischer, *Ber. Dtsch. Chem. Ger.*, 1894, **27**, 2985.

60. J.-L. Reymond, K. D. Janda and R. A. Lerner, *Angew. Chem., Int. Ed. Engl.*, 1991, **30**, 1711.

61. J. Yu, L. C. Hsieh, L. Kochersperger, S. Yonkovich, J. C. Stephans, M. A. Gallop and P. G. Schultz, *Angew. Chem., Int. Ed. Engl.*, 1994, **33**, 339.

62. H. Suga, N. Tanimoto, A. J. Sinskey and S. Masamune, *J. Am. Chem. Soc.*, 1994, **116**, 11197.

63. D. Shabat, S. C. Sinha, J.-L. Reymond and E. Keinan, *Angew. Chem., Int. Ed. Engl.*, 1996, **35**, 2628.

64. J. Yu, S. Y. Choi, K.-D. Moon, H.-H. Chung, H. J. Youn, S. Jeong, H. Park and P. G. Schultz, *Proc. Natl. Acad. Sci. USA*, 1998, **95**, 2880.

65. G. Wulff, *Angew. Chem., Int. Ed. Engl.*, 1995, **34**, 1812.

66. L. Vial, J. K. M. Sanders and S. Otto, *New J. Chem.*, 2005, **29**, 1001.

67. (a) G. W. Huber, J. N. Chheda, C. J. Barrett and J. A. Dumesic, *Science*, 2005, **308**, 1446; (b) C. J. Barrett, J. N. Chheda, G. W. Huber and J. A. Dumesic, *Appl. Catal. B - Environ.*, 2006, **66**, 111; (c) J. N. Chheda and

J. A. Dumesic, *Catal. Today*, 2007, **123**, 59; (d) D. A. Simonetti and J. A. Dumesic, *Chem. Sus. Chem.*, 2008, **1**, 725.
68. B. List, R. A. Lerner and C. F. Barbas III, *J. Am. Chem. Soc.*, 2000, **122**, 2395.
69. Y. Hayashi, S. Aratake, T. Okano, J. Takahashi, T. Sumiya and M. Shoji, *Angew. Chem., Int. Ed. Engl.*, 2006, **45**, 5527.
70. R. K. Zeidan, S.-J. Hwang and M. E. Davis, *Angew. Chem., Int. Ed. Engl.*, 2006, **45**, 6332.
71. L. Zhong, J. Xiao and C. Li, *Chin. J. Catal.*, 2007, **28**, 673.
72. T. Setoyama, *Catal. Today*, 2006, **116**, 250; plus references therein.
73. C. A. Gaertner, J. C. Serrano-Ruiz, D. J. Braden and J. A. Dumesic, *J. Catal.*, 2009, **266**, 71.
74. K. Tanabe and T. Yamaguchi, *Catal. Today*, 1994, **20**, 185.
75. K. Tanabe and W. F. Hölderich, *Appl. Catal. A - Gen.*, 1999, **181**, 399.
76. T. Yamaguchi, H. Sasaki and K. Tanabe, *Chem. Lett.*, 1973, **9**, 1017.
77. K. Takahashi, T. Hibi, Y. Higashio and M. Araki, *Shokubai*, 1993, **35**, 12.
78. K. Tanabe, *J. Chin. Soc.-Taip.*, 1998, **45**, 597.
79. J. Macht, R. T. Carr and E. Iglesia, *J. Am. Chem. Soc.*, 2009, **131**, 6554.
80. J. Macht, R. T. Carr and E. Iglesia, *J. Catal.*, 2009, **264**, 54.
81. M. K. Kidder, P. F. Britt, A. L. Chaffee and A. C. Buchanan III, *Chem. Commun.*, 2007, **1**, 52.
82. M. K. Kidder, P. F. Britt, Z. T. Zhang, S. Dai, E. W. Hagaman, A. L. Chaffee and A. C. Buchanan III, *J. Am. Chem. Soc.*, 2005, **127**, 6353.
83. J. M. Notestein, L. R. Andrini, V. I. Kalchenko, F. G. Requejo, A. Katz and E. Iglesia, *J. Am. Chem. Soc.*, 2007, **129**, 1122.
84. J. Hine, M. S. Cholod and W. K. Chess, *J. Am. Chem. Soc.*, 1973, **95**, 4270.

Subject Index

Page numbers in *italics* refer to figures or tables.